Schriftenreihe der SRH Hochschule Heidelberg

Band 8

Schriftenreihe
der SRH Hochschule Heidelberg

Band 8

Herausgegeben von der

SRH Hochschule Heidelberg

Katja Kuhn

New Perspectives of Sustainable Management in Different Worlds

Logos Verlag Berlin

λογος

Schriftenreihe der SRH Hochschule Heidelberg

herausgegeben von der
SRH Hochschule Heidelberg
Private staatlich anerkannte Fachhochschule
Ludwig-Guttmann-Str. 6
69123 Heidelberg

www.srh.de

Bibliografische Information der Deutschen Nationalbibliothek

Die Deutsche Nationalbibliothek verzeichnet diese Publikation in der
Deutschen Nationalbibliografie; detaillierte bibliografische Daten sind
im Internet über http://dnb.d-nb.de abrufbar.

ISBN 978-3-8325-2926-0
ISSN 1866-9034

Umschlaggestaltung: BüroB, Barbara Leibig
bl@buerob.de, www.buerob.de

Logos Verlag Berlin GmbH
Comeniushof, Gubener Str. 47,
10243 Berlin
Tel.: +49 (0)30 / 42 85 10 90
Fax: +49 (0)30 / 42 85 10 92
http://www.logos-verlag.de

CONTENTS

RECENT DEVELOPMENTS ON LEGISLATION AND POLICY WITH REGARDS TO BROWNFIELDS MANAGEMENT IN MALAYSIA

Chun-Yang Yin[1] and Suhaimi Abdul-Talib[2]

Universiti Teknologi MARA, Shah Alam, 40450 Selangor, Malaysia
[1]Faculty of Chemical Engineering
[2]Faculty of Civil Engineering
Environmental Engineering Technical Division, The Institution of Engineers, Malaysia
Email: *ecsuhaimi@salam.uitm.edu.my*

ABSTRACT

This paper highlights the legislations, policies as well as development of guidelines with regards to brownfields management and remediation of contaminated land in Malaysia. The legislations, policies and guidelines on such sites, although virtually non-existent in the 80s and 90s, have been comprehensively formulated by governmental agencies in recent times and currently being implemented. The comprehensive nature of such legislations, policies and guidelines indicates that Malaysia is moving in the right direction in terms of brownfields management and governance.

INTRODUCTION

The presence of brownfields is of concern in most developed countries due to scarcity of uncontaminated and arable land for new development as well as public health concerns. Remediation and/or reclamation of such sites are highly desired as such undertakings can enable sustainable development of built-up areas. In Malaysia, brownfields are found at many locations such as motor workshops, petrol stations, fuel oil depots, railway yards, bus depots, waste disposal sites, landfills, ex-mining lands, industrial sites and sites with underground storage tanks. Many of the older industrial areas in Malaysia have large patches of contaminated land. Most of these older sites have now become very much a part of the inner city areas; areas with high land value and development potential.

With the increasing population and the pressure for a more sustainable urban development, the practice of going further from the city centre to »greenfield« areas must be reviewed. Some of the brownfield sites located within or on the peripheral of the city centre in Kuala Lumpur have been redeveloped. Amongst the high profiled developments are the Mines Resort, the Sunway Pyramid Resort and the KL Central Station. Considering that Malaysia does not have any specific land contamination laws requiring soil cleanups before brownfields are redeveloped, these redevelopment projects are commendable. These projects were carried out by companies that have in house expertise using accepted guidelines established in other countries. In recent times, relevant stakeholders have recently begun to be aware of the problems associated with contaminated and are now beginning to formulate legislations and policies in a holistic manner to address this growing dilemma. This paper highlights recent developments on legislation and policy with regards to brownfields management in Malaysia. Development of guidelines and standards in this regards is also reported.

ENVIRONMENTAL QUALITY ACT 1974

Legislations with regards to contaminated land in Malaysia began with enactment of the Environmental Quality Act, 1974 which was amended in 1996 (hereafter referred to as EQA 1974). Numerous regulations concerning water, air and noise pollution are promulgated under EQA 1974 but it lacks specific legislation governing contaminated land (Balamurugan and Victor, 2001). Soil contamination is concisely stated in Section 24 of EQA 1974 and is addressed as follows:

(i) No person shall, unless licensed, pollute or cause or permit to be polluted any soil or surface of any land in contravention of the acceptable conditions specified under Section 21.

(ii) Not withstanding the generality of subsection (i), a person shall be deemed to pollute any soil or surface of any land if:

(a) he places in or any soil in any place where it may gain access to any soil any matter whether liquid, solid or gaseous; or

(b) he establishes on any land a refuse dump, garbage pit, soil and rock disposal site, sludge deposit site, waste injection well or otherwise used land for the disposal of a repository for solid or liquid wastes so as to be obnoxious to human beings or interfere with underground water or be detrimental to any beneficial use of the soil or the surface of the land.

They general nature of Section 24 of the EQA 1974 has significant conse-quences on subsurface contamination in Malaysia as the absence of spe-cific legislation enable polluters to absolve themselves of any blame in the event of a discovered contaminated land. In fact, there are many errant lo-cal factory operators that resort to illegal burial of generated toxic and haz-ardous wastes within their premises as well as dumping of such wastes in secluded plantation areas in order to save on their treatment costs. Such sites are subsequently abandoned and become derelict in due time.

ENVIRONMENTAL QUALITY (SCHEDULED WASTES) REGULATIONS 2005

The *Environmental Quality (Scheduled Wastes) Regulations 2005* (EQA (SW) R 2005) came into force on 15 August 2005 aimed to provide a more focused and better governance on scheduled wastes classification, generation, transportation and treatment. In Malaysia, the term »scheduled wastes« is used to represent the broad range of hazardous wastes listed in respective hazardous wastes-related regulations (Yin et al., 2007). Realizing the importance of a specific legislation on soil contamination to circumvent loopholes which existed in EQA 1974, the DOE included a special provision on contaminated soil in EQA (SW) R 2005. The Regulations classify contaminated soil, debris or matter resulting from clean-up of a spill of chemicals as scheduled wastes which require treatment or disposal at prescribed premises (Lee, 2006). More specifically, contaminated soil is listed in Group 4 (wastes which may contain either inorganic or organic constituents) SW 408 as *contaminated soil, debris or matter resulting from cleaning-up of a spill of chemical, mineral oil or scheduled wastes*. In addition, Regulation 14 (Spill or Accidental Discharge) of EQA (SW) R 2005 clearly specifies the role of waste generator should there be any spill or accidental discharge into the environment. This, of course, includes spillage of chemicals onto a particular site. The full statements of Regulation 14 (Spill or Accidental Discharge) are listed in the following:

(1)   In the event of any spill or accidental discharge of any scheduled wastes, the contractor responsible for the waste shall immediately inform the Director General of the occurrence.

(2)   The contractor shall do everything that is practicable to contain, cleanse or abate the spill or accidental discharge and to recover substances involved in the spill or accidental discharge.

(3)   The waste generator shall provide technical expertise and supporting assistance in any clean-up operation referred to in subregulation (2).

(4)   The contractor shall undertake studies to determine the impact of the spillage or accidental discharge on the environment over a period of time to be determined by the Director General.

Of interest is item (3) which stipulate that the polluter must provide clean-up experts to conduct remedial activities. As such, these new legislations directly obligate soil polluters to clean up contaminated soil either via *in-situ* or *ex-situ* methods.

IMPLICATIONS OF EQA (SW) R 2005

With the enactment of EQA (SW) R 2005, the local industrial backdrop will be significantly affected in one way or the other. This directly results in a more significant role for contaminated soil remediation experts whose previous related undertakings were limited to remediation projects borne by large multinational corporations. This is because EQA (SW) R 2005 legally compels soil polluters (be they small and medium industries or large corporations) to treat contaminated soil as a result of spilled chemicals even though contamination occurs within their own industrial premises. As such, services of soil remediation experts will be highly sought after. EQA (SW) R 2005 may also have a positive impact on local sustainability initiatives in terms of recycling of e-wastes. The regulations shall provide the impetus to various stakeholders to conduct more R&D initiatives to improve on treatment, recycling and reuse of contaminated soils and e-waste.

POLICIES, GUIDELINES AND STANDARDS

The National Urbanization Policy (NUP) 2006 which was formulated by the Federal Town and Country Planning includes key policies that are relevant to soil remediation (Hashim, 2006). Policy number 6 of the NUP calls for urban development as a priority for development strategies in urban areas which includes the following strategies:

    a.   Implementation of infill development in potential areas.
    b.   Identification and registration of contaminated land.
    c.   Rehabilitation of contaminated sites prior to being developed.

d. Promotion of private sector's involvement in urban redevelopment by provision of incentives and joint ventures with government agencies.

The Department of Environment (DOE), Malaysia is currently engaging contaminated soil remediation and brownfields experts and consultants to formulate a more comprehensive set of legislations to further bolster effective management of such sites. To facilitate this effort, similar local regulations as well as legislations from other countries are to be extensively reviewed. Local legislations which can be used for the purpose of managing contaminated land are:

a. Environmental Protection Enactment (Sabah),
b. Geological Survey Act 1974,
c. Land Code (Peninsular Malaysia),
d. Land Code 1957 (Sarawak),
e. Land Ordinance (Sabah),
f. Mineral Development Act 1994,
g. Occupational Safety and Health Act,
h. Town and Country Planning Act (Peninsular Malaysia),
i. Town and Regional Planning Enactment (Sabah).

Under the Ninth Malaysia Plan (2006–2010), DOE, Malaysia has initiated studies on criteria and standards for managing and rehabilitation and remediation of derelict and contaminated land. The purpose of these initiatives is to ensure consistent procedures for assessment and subsequent restoration of contaminated sites (Ghazali, 2006). Guidelines that have been proposed include:

a. Guidelines for Assessing and Reporting Contaminated Sites,
b. Guidelines for the Remediation of Contaminated Sites,
c. Guidelines for the Planning and Management of Contaminated Land.

In a related but separate development, Standards and Industrial Research

Institute of Malaysia (SIRIM) a corporatized entity under the Ministry of Science, Technology and Innovation, had established Working Groups to draft standards related to the contaminated site remediation. A draft of site remediation guideline entitled »*Developing and Implementing Early Action Guidelines for Site Remediation*« is currently being developed and available for public comment soon. This guideline provides guidance for assisting in the development, selection, design, and implementation of partial, short-term, or early action remedies undertaken at sites of waste contamination for the purpose of managing, controlling, or reducing risk posed by environmental site contamination. Early action remedies and strategies are applicable to the management of other regulatory processes (e.g. *Solid Waste and Public Cleansing Management Bill 2007* and other pertinent regulations).

CONCLUDING REMARKS

The comprehensive nature of the above-mentioned new legislations, policies and guidelines on remediation and management of derelict and contaminated land indicates that Malaysia is moving in the right direction in terms of brownfields management and governance. It is the hope of the authors that a local legislation similar to the *Superfund* and CERCLA legislations (in the United States) be enacted in the future so that there is a direct and standalone legislation that specifically aims to prevent creation of contaminated sites and comprehensively governs remedial actions. This is to ensure the existence of strict regulations which can facilitate safe and effective contaminated land clean-ups in Malaysia.

REFERENCES

Balamurugan, G. and Victor, D. J. (2001),
The applicability of ASTM E1527 standard for phase one site assessments in Malaysia, Proceedings of the National Conference on Contaminated Land: Brownfield, 14–15 February 2001, Petaling Jaya, Selangor, Malaysia.

Ghazali, H. (2006),
Contaminated land management in Malaysia,
IMPAK, 3, 14–15.

Hashim, N. (2006),
Towards sustainable development of brownfields in Malaysia,
Keynote lecture presented at Brownfield Asia 2006: International
Conference on Remediation and Management of Contaminated Land,
5–7 September 2006, Kuala Lumpur, Malaysia.

Lee, H. K. (2006),
Scheduled waste management: Issues and challenges,
IMPAK, 2, 4–13.

Yin, C. Y., Abdul-Talib, S., Lee, H. K. (2007),
*Current scenario and future directions in scheduled waste management
in Malaysia,* Proceedings of the 25th Conference of ASEAN Federation
of Engineering Organizations, November 2007, Cebu, Phillipines.

ADVANCED REMEDIATION TECHNOLOGIES (ARTS) OPTIMISING CRUDE OIL CONTAMINATED SOIL BIOREMEDIATION

Dr.-Ing. Shahrokh Peykarjou

Senior Project/Program Manager, Senior Research Scientist
Email: *Sh.peykarjou@ibl-umweltfactory.de, sh.peykarjou@ett-umweltfactory.com*
IBL Umwelt- und Biotechnik GmbH
ET&T- Environmental Training and Transfer GmbH
Wieblinger Weg 21, 69123 Heidelberg, Germany

SUMMARY

This scientific article presents in three individual phases based on analysing bioactivity and enzymatic capabilities of consortium microbial indigenous for crude oil contaminated soil bioremediation.

The designed system was multi-function, slurry phase combined bioreactors. A batch operation system with aerobic/anaerobic alternative conditions for both fixed-bed growth and/or suspended-growth multi-operation processes.

Results are achieved from 300 soil samples, each 1,200 g that sampled from 20 locations within an area about 10 hectare and the depth 0–25 (30) cm. Sampling, on-site analysing, preservation, storage and lab analysing followed with standard methods and related procedures. These locations were previously low oil contaminated sites, vegetation/farming lands, waste and sludge deposit locations and industrial sites.

Analyses followed by isolating, enhancing and separating TPH degraders microbial consortium from soil samples to reach the best bioactive and high potential microorganisms which adapted to local environment. Based on the results the soil from the previously low oil contaminated site was approved to have the highest potential and capabilities for crude oil (TPH) bioremediation. Therefore this soil was selected to be used during other laboratory analytical tests and experiments.

The selected soil with the highest biodegradability was used for group of tests to optimize essential environmental parameters for crude oil bioremediation under designed system. By these lab analysing experiments the optimum levels and conditions based on selected soil sample and its contaminant was achieved.

These parameters were main nutrients N and P (types, concentration and ratio), pH level, temperature, salinity and contamination concentration. Also the evaporation rate, oxygen and moisture concentration were determined and optimized.

The main selected monitoring factors during analysing test and operation processes were: Crude oil residue, disappearance of crude oil, CO_2 evolution, O_2 utilisation, change in microbial density/time in coloning forming units (CFU's) and enzymatic tests whenever was needed. Some gases such as CH_4 were also measured.

To compare different systems the individual ex-situ slurry-phase biodegradation in suspended-growth and fixed-bed growth in aerobic and/or anaerobic condition were also designed and operated (it was practically approved that the combined system has higher efficiency and capabilities for biodegradation of selected crude oil contaminated soil).

To approve the achieved results a semi-field scale bioremediation for crude oil contaminated soil (10,000 ppm TPH) was also tested under a batch designed system at optimised environmental levels and conditions. System was successfully operated and the soil TPH concentration was reduced to > 98% (< 100 ppm after 60 days operation).

According to this scientific research, employing adapted indigenous microbial consortium under multi-function, slurry-phase, ex-situ bioreactor system at optimized envirnmental conditions is a high efficiency, fast and safe technology for crude oil contaminated soil bioremediation as a selected approved system, therefore:

- The consortium indigenous microorganisms in the crude oil contaminated soil from previously polluted locations (adapted local microbial consortium) have a better capability to biodegrade a higher % and wider range of crude oil hydrocarbons.

- Operating bioremediation at optimized and controlled environment during the whole processes, provides the best conditions for enhancing consortium indigenous microorganisms capabilities for microbial degradation and producing suitable biosurfactant (qualitative & quantative) by selected microbial collection.

- Bioremediation of crude oil contaminated soil through employing indigenous microbial consortium from the same or similar previously oil contaminated at optimum environmental conditions is a safe, fast, practical and cost effective method based on selected system.

GOALS

- Experimental laboratory-scale and semi-field scale analysing to determine biodegradability and capability of adapted indigenous microbial consortium for crude oil contaminated soil biodegradation.

- Optimizing main environmental parameters to enhance selected microbial consortium bioactivities for oil contaminated soil bioremediation.

- Enhancing bioremediation processes through increasing microbial growth, reproduction and activities by releasing the enzymes (type and concentration) at optimized environmental conditions.

- Comparing and evaluation of suspended-growth, fixed-bed growth, and combination of suspended-growth & fixed-bed growth bioreactors efficiency for crude oil contaminated soil biodegradation, under slurry-phase system.

PHASE-I MAIN GOALS

Laboratory bio-testing experiments to determine the highest bioactivity level and capabilities in sampled soils to determin consortium indigenous microorganisms with the highest quantity and quality potential for crude oil contaminated soil biodegradation.

PHASE-II MAIN GOALS

Experimental laboratory analysing to optimise the environmental essential parameters at proper level, type, concentrations, ratio and conditions.

PHASE-II EXPERIMENTAL ANALYSING

(1) Analysing crude oil evaporation rates under natural and lab conditions: Type, composition and concentration considerations.

(2) Analysing soil specifications and its main characteristics: Type, texture, pH level, T, specific gravity, water and air contents, loss-on-ignition, TOC, P, N and biodegradability rate etc.

(3) Analysing the best suitable slurry-phase combination: Slurry contains from 10% up to 40% solids by weight (pre-analysing optimization).

(4)   Optimising temperature range from 4°C to 60°C with scheduled 2–3 additional degree in 10 rates.

(5)   Optimisation pH level with selected ranges from 0 to 14 in 12 scheduled forms with 2–3 additional levels to optimize pH range.

(6)   Optimisation oxygen concentration supply: Oxygen was provided for aerobic condition, through air supplying at 0.1–0.5 l/minute, for anaerobic condition, the aeration was programmed to be stopped for certain calculated time whenever was needed (pre-analysing optimization).

(7)   Optimisation nutrients type, concentration and ratio C:N:P: Effect of additional Phosphorus, Nitrogen, Nitrogen-and-Phosphorus, Nitrogen-or-Phosphorus, optimisation of C:N:P ratio, optimising the nutrients type and the levels (C:N:P:K:S tests were also analysed).

(8)   Fixed-bed material: Introducing a suitable low-cost material which can provide safe, suitable and large surface for microbiological growth and activity. Providing a better contact between microbial community, contaminated soil particles and essential nutrients, through fixed-bed material. Comparison of suspended-growth, fixed-bed growth, and combination of suspended-growth and fixed-bed growth in bioreactors for evaluating the efficiency and characteristics of each system.

(9)   Optimisation of salinity: Analysing the effect of additional NaCl as inhibitor or accelerator and optimising salinity % in crude oil contaminated soil bioremediation in presence of addition of 0% (as control sample) up to 2.0% NaCl.

(10) Optimisation of Crude Oil Contamination Concentration: Sample with additional of 0.0% crude oil selected as control sample and samples with additional crude oil up to 0.9% w/w

were prepared. The samples were tested for analysing maximum contaminant concentration capacity with no negative impact in crude oil contaminated soil bioremediation under optimized conditions (pre-analysing optimization).

PHASE-III MAIN GOALS

Operating semi-field scale biological remediation for selected crude oil contaminated soil under optimisation of all essential environmental parameters.

The designed system was consists of combined bioreactors, which can operated individually or in combination forms, where all the essential parameters can be controlled at optimum levels during the whole treatment processes. The system was operational with high capabilities for aerobic/anaerobic, suspended-growth and fixed-bed growth conditions, while all essential environmental parameters were optimised during the whole microbial processes and operation.

Controlling the total essential parameters in the same place and at appropriate time in a homogeny form provides the best condition for a complete high efficiency bioremediation with all capabilities, which are necessary for a high rate bioremediation.

In this system the solids are kept in direct contact with nutrients (type, ratio and concentration) while microorganisms are growing and active in suspended-growth, fixed-bed growth or combination of both forms in bioreactor/s.

In aerobic condition oxygen could be supplied through air (0.1 – 0.5 litre air/minute), and pH can be controlled by adding acid or alkali, usually in forms HCl (1 M) and NaOH (1 M), whenever be necessary.

Microorganisms may be also added if a suitable population is not presented, specially when due to high contamination concentration, the indigenous microbial consortium diversity or/and population do not be sufficient.

Type and Specification of Contamination

The same crude oil type in contaminated soil was used for laboratory analysing prepared samples. Some of crude oil specifications are such as: Carbon 84.6% (80%–87%), Hydrogen 12.8% (10%–14%), Oxygen 0.6% (0.05%–1.5%), Water Content 0.025% Vol, Nitrogen 0.1% (0.1%–3%), Sulfur 1.5% (1.0%–1.7%), Vanadium 61.0 ppm, Lead 1.0 ppm, Nickel 20.0 ppm, Iron 4–5 ppm and Total Acidity 0.05 mg KOH/g.

Soil Sampling and Soil Preparation

The total sampling area was around 10 hectares, which were divided to 20 locations, from each 0,5 hectare of each location 15 soil samples were sampled according to soil sampling standard method.

Based on pre-analysing tests 0–25 (30) cm layer of soil was the most important depth due to site specification and goals: contamination concentration, microbial activity, and the highest electrical conductivity for soluble sodium. Whenever the dried soil samples were needed, due to analyzing type, the samples were air-dried, or were drying in an oven at $T = 105°C$ to constant weight. Drying and weighting cycle repeated until successive weightings differ by no more than 1–2 mg.

The soil samples were also mechanically sieved (preparing 2 mm soil particles and keeping in desiccator for analyzing tests). Bulk density, specific gravity, loss-on-ignition and more parameters were also analysed.

Some determinations such as soil moisture content, pH and soil temperature have been also analysed at the sampling locations or local laboratory.

Key Indicators for Suitable Soil Selection

The key indicators were CFU's, crude oil residue %, CO_2 evolution, enzymatic activities (whenever was needed) and O_2 (electron acceptors) consumption. All the experimental laboratory analyzing approved that diversity and the abundance of microorganisms in soil from previous oily polluted area with acclimated microorganisms to crude oil hydrocarbons (sufficient repeating exposure usually in duration of more than 5 years) had the most suitable source of adapted indigenous microbial consortium.

Selected soil with acceptable bioegradability and its specifications:

TPH: 2,000 ppm, soil pH (d.water): 7.61, specific gravity: 1.18, TOC-%: 0.2 twt of soil, CFUs (g dried soil): 6.104, soil type: slightly alkaline, nitrogen g/kg soil: 0.5, phosphorus mg/kg soil: 2.4, loss-on-ignition %: 2.20, water content %:1.40, biodegradation rate: high (acceptable), soil texture: sandy loam, E.C. ds/m: 3.4, bulk density g/cm3: 1.19, soil temperature: 28°C (at sampling time).

Laboratory Experiment – The Effect of Additional Phosphorus in Crude Oil Contaminated Soil Bioremediation

In lab experiments the following combination was used: mineal medium: 100 ml, microbial solution: 25 ml, contaminated soil: 50 g, distilled water: 125 ml, number of samples: 5.

- Sample-1: Control sample;
- Sample-2: Sample-1 + 25% K_2HPO_4;
- Sample-3: Sample-1 + 50% K_2HPO_4;
- Sample-4: Sample-1 + 75% K_2HPO_4;
- Sample-5: Sample-1 + 100% K_2HPO_4.

According to laboratory experimental results sample-4 with 75% additional phosphorus as K_2HPO_4 had the optimum phosphorus concentration for selected crude oil contaminated soil sample bioremediation.

Laboratory Experiment – The Effect of Additional Nitrogen in Crude Oil Contaminated Soil Bioremediation

Analysing suitable Nitrogen source (with 50% additional of nitrogen as a result of pre-analysing). Number of samples: 5.

- Sample-1: Control sample;
- Sample-2: Sample-1 + 50% NH_4NO_3;
- Sample-3: Sample-1 + 50% NH_4Cl;
- Sample-4: Sample-1 + 50% KNO_3;
- Sample-5: Sample-1 + 50% $(NH_2)_2CO$.

Due to experimental laboratory analysing results, sample-5 $(NH_2)_2CO$ and sample-3 NH_4Cl, which contain additional 50% nitrogen had the best results. In comparison $(NH_2)_2CO$ was preferred.

Laboratory Experiment – The Effect of Additional Nitrogen and Phosphorus (individually or in combination) in Crude Oil Contaminated Soil Bioremediation

The additional nitrogen in form NH_4Cl and phosphorus in form K_2HPO_4 were added to samples-2, -3 and -4 as:

- Sample-1: Control sample;
- Sample-2: Sample-1 + 75% K_2HPO_4;
- Sample-3: Sample-1 + 50% NH_4Cl;
- Sample-4: Sample-1 + 50% NH_4Cl and 75% K_2HPO_4.

According to laboratory experimental analysing results, sample-4 includes 50% additional nitrogen in form of NH_4Cl, and 75% additional phosphorus in form of K_2HPO_4 had the optimum nitrogen and phosphorus. This experiment approved both type, and percentage of two main nutrients (N & P) had direct effects in bioremediation rate through microbiological activities. $(NH_2)_2CO$ can be also used instead of NH_4Cl as the Nitrogen source.

Laboratory Experiment – The Effect of Additional Nitrogen or Phosphorus Comparison in Crude Oil Contaminated Soil Bioremediation

- Sample-1: Control Sample;
- Sample-2: 75% K_2HPO_4;
- Sample-3: 50% NH_4Cl.

According to all laboratory experimental analysing the two main nutrients N and P in form and concentration of 50% NH_4Cl and 75% K_2HPO_4, as optimum main nutrients were selected due to their capabilities for microbiological activity and improving bioremediation rate. The results of this experiment approved important points for enhancing bioremediation (quality & quantity) through increasing microbial metabolism, reproduction and enzymatic activities.

Laboratory Experiment – The Effect of Various Crude Oil Concentration in Crude Oil Contaminated Soil Bioremediation

- Sample-1: Control Sample;
- Sample-2: Sample-1 + 0.1 A;
- Sample-3: Sample-1 + 0.3 A;
- Sample-4: Sample-1 + 0.5 A;
- Sample-5: Sample-1 + 0.7 A;
- Sample-6: Sample-1 + 0.9% A.

(A = 10,000 TPH ppm in contaminated soil).

Crude Oil Soil Samples Concentration: Sample-1: 1,000 ppm; Sample-2: 2,000 ppm; Sample-3: 4,000 ppm; Sample-4: 6,000 ppm; Sample-5: 8,000 ppm and Sample-6: 10,000 ppm.

According to experimental laboratory analysing, in comparison of all samples, the maximum biodegradation of crude oil was related to Sample-3 with 4,000 ppm crude oil contamination (biodegradation up to 97.5%). Therefore 4,000 ppm has selected as optimum crude oil concentration in selected contaminated soil before starting bioremediation.

Laboratory Experiment – The Effect of Salinity (NaCl) Concentration in Crude Oil Contaminated Soil Bioremediation

Analysing salinity concentration in form of NaCl in crude oil contaminated soil bioremediation processes, as an inhibitor or accelerator factor and optimising salinity to enhance microbiological activity (based on total salinity and salt composition).

Additional NaCl % (with standard slurry mixture):
- Sample-1: Control Sample 0% NaCl;
- Sample-2: Sample-1 + 0.3% NaCl;
- Sample-3: Sample-1 + 0.6% NaCl;
- Sample-4: Sample-1 + 0.9% NaCl;
- Sample-5: Sample-1 + 1.2% NaCl;
- Sample-6: Sample-1 + 1.5% NaCl;
- Sample-7: Sample-1 + 2.0% NaCl.

According to soil and contaminant specifications and type of salt the maximum acceptable salinity level as NaCl was 0.9% dry soil.

Laboratory Experiment – The Effect of pH in Crude Oil Contaminated Soil Bioremediation

- Sample-1: pH = 0 – 3;
- Sample-2: pH = 3 – 5;
- Sample-3: pH = 5 – 7;
- Sample-4: pH = 7 – 9;
- Sample-5: pH = 9 – 11;
- Sample-6: pH = 11 – 14;
- Sample-7: pH = 4 – 6;
- Sample-8: pH = 6 – 8;
- Sample-9: pH = 8 – 10;
- Sample-10: pH = 4 – 7;
- Sample-11: pH = 7 – 10;
- Sample-12: pH = 5 – 9.

According to all laboratory experimental analysing the optimum pH range

was pH = 5 – 9 as the best pH for bioremediation of selected crude oil contaminated soil sample. This range of pH is also suitable for most enzymatic activities and is a common range for essential nutrients solubility
such as phosphorus.

*Laboratory Experiment – The Effect of Temperature in Crude Oil
Contaminated Soil Bioremediation*

Attention: Temperature has direct effects on contamination physical
change, chemical reaction speed, biological growth, microorganisms survival and internal and external enzymatic activities and reactions. Therefore cardinal temperature study, specially understanding the optimum
rate is so important in biological remediation processes.

- Sample-1: T = 4°C;
- Sample-2: T = 15°C;
- Sample-3: T = 20°C;
- Sample-4: T = 25°C;
- Sample-5: T = 30°C;
- Sample-6: T = 35°C;
- Sample-7: T = 40°C;
- Sample-8: T = 45°C;
- Sample-9: T = 50°C;
- Sample-10: T = 60°C.

(based on pre-analysing results).

According to all laboratory experimental analyzing and results, the optimum temperature was related to sample-6 T = 35°C ± 2°C. As providing,
keeping and operating the system at exact optimized temperature during
the whole processes specially in field-scale may cause difficulty and increase costs, therefore the range T = 20 – 35°C ± 2°C was selected as suitable temperature range.

*Laboratory Experiment – Optimisation of C:N:P Ratio in Crude Oil
Contaminated Soil Bioremediation*

The C:N:P ratio in crude oil, soil and minimal media were as: C:N in
crude oil = 846:1, C:N:P in soil sample = 800:200:1 and N:P in minimal
media = 2:3. Based on pre-analyzing results and according to the results

from optimizing additional N and P by adjusting the nitrogen and phosphorus levels with NH_4Cl and K_2HPO_4, the selected C:N:P for analyzing and finding the best ratio were:

- Control Sample: C:N:P = 100:15:3;
- Sample-1: C:N:P = 100:22.5:3;
- Sample-2: C:N:P = 100:15:5.25;
- Sample-3: C:N:P = 100:22.5:5.25;
- Sample-4: C:N:P = 100:30:6.

According to CO_2 evolution, Crude Oil residue % and CFU's results C:N:P 100:22.5:5.25 in sample-3 was selected as optimum C:N:P ratio.

Laboratory Experiment – Comparing Suspended-growth, Fixed-bed growth and Combination of Suspended-growth with Fixed-bed growth Bioreactors for Bioremediation of Crude Oil Contaminated Soil.

Comparison of suspension-growth, fixed-bed growth and combination of suspension-growth & fixed-bed growth bioreactors capabilities for enhancing crude oil contaminated soil bioremediation were done with analyzing Xylit. Xylit as a suitable large enough surface area support material for microorganisms fast growing beds, improving microbiological growth and reproduction and as a result: increasing biodegradability of microorganisms for enhancing bioremediation rate.

Experimental Condition: Slurry-phase, Ex-situ, Aerobic condition (0.5 liter air/minute)

Summary of Procedure: To each Laboratory bioreactor A1, A2, B1, B2, C1 and C2, the following materials were added. All bioreactors were operated 168 hours at T = 22 ± 2°C with mixing 100 RPM, contains special equipment as upper and lower mixers to work with aeration to improve slurry mixing. All the 3 bioreactors A1, B1, and C1 were used as control and microorganisms were treated with Sodiumazid to inhibit microbial activities.

- Nutrient solution for experiment includes: 1,000 ml contains (K_2HPO_4, NH_4Cl, KNO_3, $MgSO_4 \cdot 7 H_2O$, trace amount of $FeSO_4 \cdot 7 H_2O$ and $Ca\,Cl_2 \cdot H_2O$) and distilled water up to 2 liters (including microbial solution).

- Contaminated soil = Slurry 30% (25%) with 4,000 ppm crude oil contamination

- pH adjustment: pH = 7.0 by HCl (0.1 N) and NaOH (0.1 N)

- Xylit: Used as solid material for Fixed-bed bioreactors

- Enriched microbial solution: 250 ml prepared from selected soil for each bioreactor A2, B2, and C2.

Materials Mixture for Bioreactors

Bioreactor A1 and A2 Batch System in condition slurry-phase, aerobic suspended-growth including soil = 600 g with 4,000 ppm crude oil and 1,000 ml nutrient solution. Bioreactor B1 and B2 batch system in condition slurry-phase, aerobic fixed-bed growth including soil = 600 g with 4,000 ppm crude oil with 1,000 ml nutrient solution and Xylit = 45 g as fixed-bed material. Bioreactor C1 and C2 batch system in condition slurry-phase, aerobic combination of fixed-bed growth & suspended-growth including soil = 600 g with 4;000 ppm crude oil with 1,000 ml nutrient soloution and Xylit = 45 g as fixed-bed material.

All the laboratory experimental tests for bioreactors A1, A2, B1, B2, C1, and C2 were operated under same conditions and concentrations with same material. Operation was done, while the selected most suitable indigenous microbial consortium were presented in bioreactors.

RESULTS

According to the comparison of results, Xylit can be used as a suitable natural and low-cost Fixed-bed material for TPH contaminated soil bioremediation.

Xylit: An intermediate product of wood to coal that during transformation keeps its wood structure, and can provide large surface area as a very suitable inert support medium for microorganisms activities. Microorganisms through attaching to xylit surface and growth under both aerobic and/or anaerobic conditions, can have possibility to access to a good concentration of the nutrients and contaminants as the main essential sources for growing. This conditions provide higher reproduction, more enzymatic reactions and faster crude oil (hydrocarbons) biodegradation.

Xylit can improve the physico-chemical and biological properties in soil biodegradation. High cell immobilization on this material increases microbial number as well as their activities.

The number of cells were determined in all the 3 selection forms after 72 hours, fixed-bed growth & suspended-growth = $2,4 \times 10^8$, for fixed-bed growth = $5,4 \times 10^7$ and for Ssuspended-growth = $8,2 \times 10^5$.

Therefore laboratory experimental results approved that the microbial growth for fixed-bed is 65.85 times more than suspended condition. Although Xylit was a suitable material for fixed-bed condition but as some other natural materials may have also the similar or more advantages, more laboratory studies are needed for other substances and regarding Xylit specification and characteristics as well.

The rate of crude oil degradation in immobilized condition was 1,15 times more in comparison with free cells in suspension condition. Bioremediation rate comparison was 95.9% in combination fixed-bed growth & suspended growth, 80,8% in fixed-bed growth, and 70,0% in suspended growth condition.

Phase-III Laboratory Experiment – Semi-field scale crude oil
contaminated soil bioremediation under optimum environmental
conditions in Multi-function Slurry-phase Bioreactor System

Goal: Crude oil contaminated soil bioremediation under all achieved optimum microbiological and environmental conditions and parameters.

Experimental Remediation Method: Multi-function Slurry-phase Bioreactor System

Procedure: Total mixed substances in slurry form was shifted from mixing tank in equal amount, into two bioreactors. Bioreactors were operated under batch system in combination form aerobic/anaerobic conditions, with Fixed-bed and Suspended-growth multiphase capabilities.
The same procedure was applied to control samples. During the comparison test, sampling, analysing and adjustments were done at least once
a week, or whenever was needed to control the operation progress and
monitoring.

- *Contaminants (Type & Concentration):* 10 kg crude oil
  contaminated soil (crude oil contamination: 10,000 ppm);

- *Nutrients and Salt:* Molasses and Urea-phosphate according
  to the best selected C:N:P 100:22.5:5.25, were prepared as a
  suitable and cheaper sources for nitrogen and phosphorus and
  also NaCl (0.9%) was added;

- *Fixed-bed Material:* Xylit, 5% wt of soil;

- *Temperature Range:* Control at temperature $T = 35 \pm 2°C$
  $(20 - 35°C)$;

- *pH Range:* Control $pH = 5 - 9$, pH adjusted with HCl and NaOH
  0.1 M at $pH = 7$ in the beginning of the test and whenever was
  needed;

- *Duration of experiment:* 60 days;

- *Aeration system:* 0.5 litre air per minute for each aerobic unit;

- *Mixing system:* Mechanical mixer RPM = 200;

- *Slurry mixture:* Soil as 25% (30%) of total weight and water 75% includes other materials such as nutrients and microbial solutions;

- *Enriched microbial solution:* 500 ml (prepared from selected soil);

- *Main analysing and key indicators:* CO_2 evolution, O_2 consumption, crude oil residue % or ppm, enzymatic analysing and colony forming units (CFU's)/g dry soil.

Soil preparation: Contaminated soil concentration was reduced to optimum 4,000 ppm by mixing with low contaminated/clean soil from same locations.

Control samples: The first control sample was selected for crude oil residue % comparison. This control sample was exactly similar to homogeneous suspension substances in mixing tank, with 2 : 100 ratio, under similar operational conditions in bioreactors, with exception the microorganisms were treated with sodiumazid to inhibit microbial activities. Duration of this test was the same as test sample (60 days). The second control sample was prepared for microbial growth comparison, exactly similar to homogeneous suspension substances in mixing tank, with 2 : 100 ratio, under similar operational conditions in bioreactors.

The same procedure was followed for sample test, with exception temperature was laboratory temperature (T = 20 ± 2°C), pH = 7 at beginning, with no further control, no more additional and control for nutrients concentration, duration of this comparison was 15 days.

Results of laboratory semi-field scale crude oil contaminated soil biore-mediation:

(1) According to this research, it was approved that this system under optimum environmental conditions can remediate up to around 98% of crud oil contaminanted soil in duration of 60 days.

(2) Analysing the residue Nitrogen, Phosphorus in remediated soil approved that all concentration according to experimental results and calculation were correct and there was no significant difference in comparison nitrogen and phosphorus with initial soil (Phosphorus < 3 mg/kg soil, Nitrogen < 0.7 g/kg soil in biotreated soil). Therefore the treated soil was safe to be used for suitable further purposes.

(3) According to microbial growth curve specifications, the results from the curve belongs Test-sample in compare with Control-sample approved, high fast increasing in exponential-phase and a better result in stationary-phase, and both results increase the microbial community population and activities. Therefore high biodegradation of TPH in contaminated soil under designed system was achieved.

(4) According to all laboratory experimental pre-testing and analysing it was completely approved that in bioremediation, different steps should be followed continuously and completely, and the results from each previous experiment should be used to improve the next steps. These phases start by analysing contaminated site characteristics, sampling and continue through laboratory scale, semi-field scale analysing, and will be completed by introducing a high efficiency bioremediation method for field scale. Preparing scientific proposal for all steps can be the best procedure for achieving the optimum results in crude oil contaminated soil bioremediation.

As the main goal is to reach a high efficiency and cost effective technol-ogy for field scale remediation, therefore it is so important this goal be fol-

Analysing Contaminated Soil During Bioremediation Process in 60 days – Dried Soil

 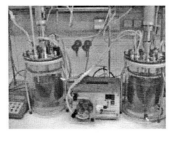 

Fixed-bed Growth Bioreactor   Combined Fixed-bed Growth & Suspended Bioreactors   Suspended Growth Bioreactor

Crude oil Contaminated Soil and Vegatation in Bioremediated Soil

> 10,000 ppm TPH   < 4,000 ppm TPH   < 2,000 ppm TPH   < 800 ppm TPH   < 100 ppm TPH

    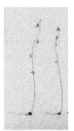

Initial Contaminated Soil

lowed in all site investigation, sampling, analysing, remediation and monitoring steps.

CONCLUSION

The results of experimental laboratory analysing approved that under optimisation of environmental conditions the crude oil hydrocarbons can be degraded by suitable consortium indigenous microorganisms at high efficient and sufficient levels. The results have also approved that crude oil contaminated soil can be bioremediated under the correct conditions mainly as:

- Using active crude oil oxidising consortium indigenous microorganisms in/from polluted sites as the most effective adapted microbial diversity and abundance.

- Introducing a suitable solid cheap waste as fixed-bed to immobilise bacteria, material or as a part of essential nutrient and/or absorbent.

- Selecting the best C:N:P ratio according to soil and crude oil properties, by providing correct concentration and type of nutrients (nutrients preferably be selected from nature or close to nature cheap materials). These substances can completely or partly be provided from waste material with considering all safety procedures.

- Providing high bioremediation rate processes through employing all optimised essential environmental parameters.

- Applying all essential materials and environmental conditions for operating designed system as a high efficiency technology which is also cost effective.

A PASSIVE: AGGRESSIVE ORGANIZATIONAL MODEL THEORY AND PRACTICE

Dr. John H. S. Craig

CEng., CEnv., Managing Consultant, FIEE, FCQI, FCMI, FIOM
John is a consultant and researcher who has spent many years in high profile senior
positions in production and was CEO of a renowned UK certification body for ten years.
He has also served many years representing the UK on EU and ISO committees.

ABSTRACT

A framework model of Passive:Aggressive Organizations is suggested by
Neilson, Pasternack and van Nuys quoting Booz Allen Hamilton as cate-
gorizing such organizations as having »quiet but tenacious resistance, in
every way but openly, to corporate directives«.

This paper tests these concepts against previous research in five large factories in Poland from 1996 to 2002 which investigated the socio-cultural
dynamics at every management level in those factories; illustrative com-
parisons are also made against these dynamics in seven large factories in
China.

It was found that, although the model explained or characterized some of
the endemic organizational dynamics, it did not do so fully, nor did their
proposed solutions identify or deal with entrenched and emergent problems contingent on those solutions, particularly as each factory had it's
own markedly different set of organizational socio-dynamics.

Finally the paper suggests some of the means, firstly to identify the underlying problems and secondly to formulate a set of measures to re-culturize the factories and remove the »tenacious resistance«.

Keywords: passive, aggressive, socio-cultural, dynamics, realities.

INTRODUCTION AND BACKGROUND

Introduction

The paper concentrates on the two pieces of research in Poland, however, in both Poland and China the relicts and actualities of the communist bureaucracy and ethos linger and, despite the different political contexts, hierarchies in the factories and their effects are surprisingly similar. In this context the paper considers the practical processes that lie underneath the generic model on passive:aggressive organizations proposed by Neilson et al., this at factory management levels, and the impact this could and does have on the formation and implementation of company strategies.

Neilson et al. (2005) propose three precepts: which create the conditions for the model for a passive: aggressive organization:

- an unclear scope of authority;
- apparent agreement (with lower management levels) without cooperation;
- misleading (organizational) goals;
- misleading (organizational) structure.

In the following discussion the model and the practice are compared using rich data from comprehensive surveys at each factory. Every manager at all levels was interviewed.

Background

The lead author was involved in the setting up, maintenance and monitor-

ing of production and other processes, quality and environmental management systems in an area of the heavy industrial sectors in Poland and China.
An inadequacy of knowledge deployment found during these examinations
of the formal management systems became an indicator for the »*command
and control*« top management techniques which were causing passive resistance at all levels down the management chain this causing a significant
lack of knowledge transfer. Some modern management methods had been
attempted in the factories but a lot of old *authoritarian* thinking remained
within strictly hierarchical structures with rigid departmental demarcations
and, sometimes, severe penalties for stepping over job boundaries.

> »*In such organizations people with authority lack the information to ex-
> ercise it wisely or the incentives to serve the company's strategy, or the
> people who will carry out the directives. As a result, many in senior po-
> sitions operate under the false impression that they control things which
> they do not.*« (Neilson et al., 2005)

The original research established that the managers formed themselves
into socio-cultural groupings based on their individual perceptions of organizational realities, aberrant linguistic interpretations of communications between management levels, and risk reduction in their jobs. From
the data, the gaps between the managers' realities and the empirical (expert benchmark) realities were determined and used to indicate the organizational dynamics and dysfunctions.

ACTION RESEARCH – SITUATIONAL METHODOLOGY

Initial research took place in the years 1998–2001 and deeper research
in the same factories[1] plus seven others in China between 2002 and 2005,
with all the managers (164) participating. This research involved charac-
terizing the socio-cultural groupings in the organizations and the effect
this was having on the organizations.

----
[1] For the initial study the heavy manufacturing factories were independent, in a
later research study they were part of the same, larger organization.

The deeper research (Craig & Lemon, 2005) into system dynamics was conducted via an administered questionnaire, rich data from the observation of manufacturing and management processes and the environmental impacts of the companies, and the benchmarking of critical impacts on production and (primarily) the environment. It also involved semi-structured one-to-one interviews with every manager focusing, in particular, on those technical issues where the lack of up-to-date process and marketing knowledge was preventing the necessary, and rapid, adjustment to the rising competition from the developed world market. These data were supplemented by notes from management, together with process and environment workshops which were primarily devoted to aspects of socio-cultural dynamics and change management (see Alasuutari, 1995 for the qualitative contexts of the research). The findings from the research were presented at various workshops and, with the exception of one senior human resources manager, were totally accepted.

Access to the research environment was facilitated through the *history* between the researcher, as a systems evaluation expert, and the companies involved. While this history was necessary to gain access it was important to recognise that such a position might be perceived in a very different way by others within a hierarchical organization, particularly where issues of trust might exist. Employees within the companies often adopted *survival* techniques that included an aversion to answering questions and an extreme reluctance to fill in forms – a type of passive resistance. Moreover, when they did fill in such forms, the answers might be politically correct and anonymous to avoid any possibility of punishment due to *singing* from a *different hymn sheet* – aggression at the top leading to risk avoidance at the lower management levels.

From the ninety four Polish managers who took part in the second piece of research, ranging from board members to supervisors, it was quite typical for people at ensuing research workshops to agree that they wanted a more open culture and improved team working but within a hierarchical structure. They were less willing to commit to this in practice however, arguing that they were too wedded to the old ways – it was for the coming generation to instigate such cultural change.

LOCATIONS AND LEADERSHIP

The Polish factories are in five industrial locations across Poland, date from the early 1900's and have been mostly engaged in the »heavy end« of industry. The environmental impact, with brown coal smoke-stack grates, from which clinker and fly ash had been spread on most of the ground (20 – 56 ha), is considerable due to contingent contamination, mainly heavy metals and toxic chemicals (with pH's ranging from 11 to 13).

The Chinese factories have a smaller footprint as they tend to have the processes concentrated on several floors of tall buildings, rather than the horizontal spread of the Polish factories. Hence any contamination is more concentrated with the open land involved being, at the most, some 1 ha, there is, however, less control on effluents to streams or rivers. The MD's or Plant Managers in China also use an authoritarian ethos but are much younger and in new factories, and are predominantly foreign Chinese (not mainland), brought up in first world economies using Western management techniques.

RULE SETS AND SOCIO-CULTURAL ABERRATIONS

It could be argued that quality and environmental management systems such as ISO14001 and ISO9001, which insist on codified rule sets, have provided a glue whereby, at least at the operational level, people can see that they are part of a unifying process. The introduction of the rule systems may provide a stabilizing, integrating force over and above their functional effectiveness. Counter-balancing this, the research established that, dependant upon topic, between seven and fourteen different socio-cultural sub-groups with divergent perceptions on these systems existed in every factory at most management levels.

Table 1:   Pattern classification (Frost et al., 1989)

country/ classification	integrated patterns %	differentiated patterns %	fragmented patterns %	number of different response patterns
Poland (5 factories)	4	4	92	56

Table 1 exhibits a classification developed by Frost et al. (1989) where »integrated« refers to groups of people with near identical perceptions of a topic or event, »differentiated« to (say) two clearly different groups, and »fragmented« to groups scattered along a perception scoring line (Likert Scale, 1932). It is clear that there is little coherent or common agreement among socio-cultural groups of managers in the five factories.

These groups occur between and within the levels of management, whereby significant differences could be identified between what top management desired and what actually happened at the worker-agency level. Here Burns and Flam (1987) suggested that people's activities may in part be strategic in character, designed in order to maintain or change rule regimes which favour their interests, or hurt the interests of others to whom they are opposed or antagonistic; (Archer 1996) goes into more detail with her analysis of how commands may be elided or conflated down management chains. These problems could be related to bureaucracy within the hierarchy, embodying a passivity occasioned by a feeling of helplessness, and dubious communication realities, but also as a delinquent methodology for creating space (Mars, 1994) to pursue one's own agenda, and less risk in decision-making (see Perugini & Bagozzi, 2001) at any particular level of management.

Table 2 indicates that, in the second set of research data, when the five independent factories where grouped under the control of a central factory HQ, there is a dichotomy between »command and obey« and »plant autonomy« as regards any independent action. The Plant Managers are obliged to be passive or they will be strongly punished (items 4 & 5).

Table 2:   Factors for comparison as cultural indicators taken from the rich data

	phenomena	factories				
		1	2	3	4	5
1	HQ command and obey ethos	Y	Y	Y	Y	Y
2	plant autonomy at factory sites	N	N	N	N	N
3	job demarcation (elision of responsibilities)	O	O	O	O	O
4	punishment culture	S	S	S	S	S
5	fear and job insecurity	Y	Y	Y	Y	Y
6	training	A	A	A	A	A
7	communication	P	P	P	P	P
8	knowledge sufficiency					
	quality in production	A	A	A	A	A
	environment	A	A	A	A	A
	health & safet	A	P	A	P	P
9	initiative & innovation	P	P	P	P	P

Codes in table: O = overt; S = strong; A = average; P = poor; Y = yes; N = no.

Both the unwelcome changes of the enterprise culture, and the resistance of managers to change that has been passed down from previous regimes are dealt with well in the literature (e.g. Cameron & Quinn, 1999; Forsyth, 2000; Thomas & Davies, 2005) and correspond closely with the situation at the factories researched. In particular this covers the key role of communication in maintaining the motivation which is so clearly lacking at the study factories. Unfortunately, the factories had no clear formal communication strategy about anything, let alone the then-current extensive down-sizing in Poland, so the amount of goodwill that remained among the lower level managers was stated, by them, as loyalty to the factory rather than to top management. This may have been grounded in long service and the feeling that this is home, despite the current inherent failings, this representing a social cohesion which is untapped for development in the realization of a coherent organizational mission (Archer, 1996) or in the human dimensions such as active motivation strategies.

COMMUNICATION IN A MISLEADING ORGANIZATIONAL CULTURE WITH MISLEADING GOALS

Figure 1 depicts the hierarchical management structure (full line) within the studied organizations, with the economic paradigms at the top. The differing feedback loops (dash dot), representing different communication and management socio-cultural groupings, came directly from the survey data (Craig & Lemon, 2004). A typical comment from the managers, in this context, was that »we get our information of what top management is doing from under the crane« – these were the cranes in the shipping yards which provided the settings for the most rapid communication paths in the factories.

However, this complex set of *touching* or linked sub-cultures (A & B & C) could be viewed as an aid to management by facilitating the translation of sometimes ambiguous or unrealistic commands into terms that were intelligible and workable for the realities of the shop floor. As the linguist Holmes (2001) says, commands must have meanings which fit into the mind-sets of the recipients, not as with a precept that *words have the meaning that I mean them to have*.

Need to know:
In an interview with two managers of their goods-in department, who reported to the managing director, the researcher questioned why the workers handling the goods did not know about the extreme dangers from some of the chemicals they were moving about; it emerged that the managers did not know either.

The Health & Safety and Environmental managers did know but took the position that neither the managers of the workers nor the workers needed to know the dangers.

The knowledge was retained in an apparent attempt to make themselves indispensable.

Figure 1: Socio-cultural interfaces within the management chain

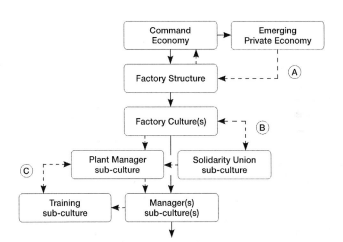

The major problem of communication was corroborated in the training workshops and was exemplified by the fact that only one or two out of over one hundred participants knew of a strategy or mission for their company. This is an indication of Neilson's (2005) organizational passivity as well as of poor communication and authoritarian management styles which were denying »worker« participation (see also Skrupinska, 2004). Information transfer was restricted to the issuing of orders as discussed above and a »need to know« philosophy, the informal transfer between workers in operational contexts, and that issued via the Solidarity Union (see Figure 1).

The situation within the Polish Group which took over and was rationalizing the previous five individual companies, and where the new senior management (the *New Blood* of Neilson's change proposals) was parachuted into large production units overnight to replace the existing MD, is even more complex. Few would commit to knowing what was happening at Board level or was likely to happen – as mentioned by Neilson, the symptoms of institutionalized dysfunction. In a personal context the researcher developed a rapport with managers and directors who might, without warning, be moved elsewhere hence leaving the *job* of re-culturization unfinished or seriously interrupted.

Such uncertainty provides a considerable constraint on the implementation and improvement of management systems. Top management may even manipulate uncertainty in order to retain control by preventing the implementation of changes that might threaten the existing order in higher management (see also Perrow, 1972). The use of parachutists to instigate and manage cultural change might, paradoxically, be preventing it.

Within the lower management the new parachutists are often perceived as *good guys*, sometimes with *acceptable* new views, who were trying to super-impose an un-synthesized nebulous culture in the alien environment, but who neglected the fact that there was an existing set of socio-cultural groups which was unlikely to be changed by coercion. Discussion with top management indicated two clear impediments within this coercive change, inadequate formulation of what the new culture should be, and no adequate training on how to recognize, and manage, the elements, mechanisms and dynamics of existing cultures which were creating opportunities for action or, as top management perceived, havoc.

From discussion with members of sub-ordinate managerial groups it was quite apparent that they often avoided taking action due to risk avoidance, and could be regarded as passively awaiting whatever fate might bring their way. In one case, rather than go out on to the production lines to assess the health and safety risk to operators and to ensure alterations to production methods, the health and safety manager, who had the authority to take decisions, formed a committee of eleven people to make those decisions – goals thereby become diffuse and misleading in the sense that the action from these decisions might not be in a top-management desired direction.

To summarize, the post-socialist process of management in this Polish industry since 1990 had caused little change to the management ethos.

The management ethos in the Chinese factories, despite being new is strangely the same as in Poland and apparently derives from the old Chinese bureaucratic (Confucian, 4th BC) legacy as against that of the European Max Weber (1922). These management processes were producing resistance to change at subordinate levels and were truly symptomatic of passive:aggressive organizations (in the Neilson sense).

HIERARCHY, JOB DESCRIPTIONS AND DEMARCATION AND AN UNCLEAR SCOPE OF AUTHORITY

The current hierarchical systems apparently formed a closed environment, and were accepted as a norm by top management. But, due to the levels of risk avoidance at lower levels mentioned above, their effectiveness slowly deteriorated through a gradual abandonment and atrophy in the use and relevance of knowledge. As Morgan (1986) says, this will often go unrecognized until too late:

> »It is no accident that organizations learn poorly. The way they are designed and managed, the way people's jobs are defined, and most importantly, the way they are taught to think and interact create fundamental learning disabilities.«

Under communism, job descriptions and responsibilities were narrowly defined with the (cultural lag) effect that it still remains difficult to get people to step outside of departmental boundaries. In this sense it can be said that job descriptions constrain horizons (Drucker, 1992) and perhaps the amount of knowledge a person perceives to be necessary in undertaking tasks or implementing change – an ambiguous scope of authority. Various writers have dwelt on the importance of job descriptions, e.g. Forsyth (2000), but may be on dubious ground as there are others who think the opposite, such as Merton (1967) and Drucker (1992), who describe the rigidity of rules and job descriptions as deeply ingrained habits which serve organizations badly. This may be a case of the structure determining the rules, which then generate consequential organizational behaviour problems.

However, in the factory contexts such restriction and rigidity may come not only from the endemic authoritarian command culture but from the mental models suggested by Patten (1996) where the personal mental irrationalities impede a sound approach to decision making. These irrationalities can mutate the original goals of the organization and mislead the top management on what is actually happening at the workface.

This affect was confirmed in research papers by Craig & Lemon (2004, 2008). There are also the *mind viruses* of Goodenough (1995) whereby *mental microbes* from a small initial group grow into a factory culture which tends to reinforce itself and infect new people. A typical example of this occurred when, as a result of rationalization, some of the training staff in Poland were relocated to a central HQ. The HQ staff had a perception that the rationalization was not having a detrimental effect on staff morale, quite the opposite of the relocated staff, but this perception became the norm for the relocated people – but significantly erroneous. Vroom (1959) has described these types of situation as having:

> »...*a syndrome of dogmatism, misanthropy and xenophobia associated with authoritarianism and low interpersonal trust*«.

In this study a significant proportion of people stated a desire for team working, but with a strong (authoritarian) leader – the corollary of this might be that an authoritarian regime attracts people with low trust who will not be open to sharing information and knowledge (Vroom, op. cit.) and who will lead the organization along dubious paths.

The situation in the surveyed factories corresponded to Townsend's (1970) view that we tend to meet any new situation by re-organizing; this can create the illusion of progress while producing confusion, inefficiency and demoralization. The use of short-term, high-level managers, the parachutists, has already been identified as one example of this and may be congruent with the feelings of many managers on the topic of successive management system failures, as Ramsey (1996) opines:

> »...poorly thought out, half-heartedly applied, but often with lip-service to unmitigated management commitment and a call for an attendant ›complete culture change in the organization‹, and then frequently abandoned in disillusion – possibly with an atmosphere of blame and recrimination.«

Using the frameworks of Archer (1996), Frost et al. (1989) for analysis, it emerged from the data that there are distinctive sub-cultures at different management levels and this appeared to be an hierarchical structural effect rather than people driven. The implications of these observations are that the *hierarchy as an entity* in itself promotes or restricts knowledge, especially if top management is unaware of the problem and managers at different levels deploy differing social constructs; these can evolve from being inexplicit differentiations to form more explicit differentiated constructs (Hegel, in Morrison, 1995) and turn into differentiated cultures at the management level with boundaries between them (Figure 2, p. 49) giving a mix of adaptive behaviour and intransigence (Neilson, 2005).

DECISIONS WITHIN HIERARCHICAL CONSTRAINTS AND AGREEMENT WITHOUT CO-OPERATION

As indicated above, if the reasoning related to the perception of a problem is in a language not shared by different levels of management, communication of any related decision may be at variance with the intention of that decision. There can be a close relationship between language and perceptions of meaning, but does language constrain perceptions or vice versa? Is it possible to think outside of that framework of cognitions? Here Heugens (2005) argues that:

> »Human behaviour is often intendedly rational, but only boundedly so. People root their decisions in subjectively derived cognitive models, which diverge among individuals, and seldomly converge due to the incompleteness of available information.«

In the Polish factories, even where people had the requisite knowledge for taking a good decision, a significant proportion geared those decisions for self-benefit. Others talked of the dangers of exhibiting knowledge, ostensibly because the manager got rid of the person if they feared competition, but perhaps also because they would then be expected to take decisions which, however inadvertently, might lead to punishment. Here, the subordinate managers were creating the conditions for an operational passivity in order to mitigate the effects of senior management aggressiveness and, perhaps, hide under an unclear scope of authority (Neilson et al., 2005).

The research data confirmed a proposition that bureaucracy stifles innovation (Merton, 1967) and there is clear evidence that the protective screen of rigid and *narrowed* job descriptions ensured that knowledge was preserved within a person and/or within a department. This knowledge was rarely used as a basis for decisions, which were only made under some form of formal or informal duress, and could not be pushed *upstairs* for someone else to take responsibility. Moreover, as a large proportion of decisions were made for the benefit of people inside, or organizations outside, the factory, this indicates socio-cultural fragmentation inside the factory for work groups, and socio-cultural integration outside it with peer and community groups – and split loyalties.

DOWN-SIZING, KNOWLEDGE AND MORALE

In the context of down-sizing, there were rapid demotions and promotions at all levels in the study factories and the managers believed that success in a hierarchy depended on the capriciousness of the manager, this leading to a fairly general cynicism. A poignant example is that, when people were promoted to another (remote) factory in a group, they always maintained their original home, *knowing* that they would need it fairly soon after the promotion. In this context Clegg and Palmer (1996) proposed that:

»Western' managers extended the warm embrace of the enterprise culture, but then many in the developing economies found that capital's ultimate embrace wrung them, very warmly, by the neck.«

It has been seen that with down-sizing, people are stretched; with the lack of communication, they are disillusioned and with a perceived failure of the market economy to bring social benefit, they are cynical. It has been suggested that people may put up with such change without liking it and, in consequence, the process may cause resentment and demotivation as noted by Forsyth (2000), an apparent passivity. In this situation an attribute of *close-mindedness* may well be killing the urge to acquire and bank knowledge for the benefit of both factory and society. Top management had, apparently, not considered the human effect of aggression on promoting the company goals, whereby the goals were elided or conflated at the lower management levels.

It is salutary that, even taking into account the necessity of a promise of strict confidentiality given to the interviewees, they were amazingly frank in their views on this subject – they appeared to need to vocalize their problems in order to relieve their feelings and the stress that resulted from them. The researcher was requested frequently for an unattributed comment to be fed to the Board. This indicates that operational (job) passivity does not preclude becoming active, in the right circumstances, to try to break out of that passiveness.

The phenomena in the Chinese factories are rather different. There is a strategy of not training people except for the minimum attributes that are required for a de-skilled job in order that the staff do not ask for consequential wage increases. In individual discussions the women said that they felt very restricted by this and, in the bigger conurbations, actively sought opportunities elsewhere. The men concentrated on looking for a higher salary but very rarely engaged in outside training. In various assessments of production methods it was found that, as a result of the paucity of training and the discouragement of innovation, production quality was suffering. The top management concentrated on production output, aggressively, but the sub-ordinate management concentrated on salary and survival rather than innovation – in this sense they appeared passive to the top management in their reaction to a requirement for better productivity.

KNOWLEDGE AND ILLUSIONS

In the above context, several writers (e.g. Sperber, 1996; Archer, 1996; Fritz, 1999) argue that people may substitute a concept of reality for reality, then impose this construction on themselves in a self-reinforcing process that results in them only seeing what they want to see. This can lead to the acting out of imaginary realities at variance with factory objectives. One concomitant illusion is that the managers felt (as demonstrated by their answers to the questionnaire) up-to-date in their knowledge of environmental legislation and best practice e.g. on groundwater contamination whereas, in practice, the knowledge level was very poor.

This perception resulted in a top-management confidence that everything was under control, this being reinforced by the norms of the socio-cultural groups identified in the data, to the extent that it provided a feeling of security within the factory and reduced or removed the imperative to acquire more knowledge and thereby avoided action on misleading, irrational, goals. Moreover, the lack of necessary knowledge was evidenced *on the ground by the researcher* and reinforces the axiom that management does not know what it does not know.

The above mentioned illusions raise the possibility that, with virtually powerless employees, knowledge gaps can be used as a way of manipulating relationships through competition, revenge etc. By extension the effect of restricting work activity to the job specification (or less) and putting other energies into family or outside life as a counterbalance to frustration at work may not differ significantly from the current *quality of life* and home/work balance debates in the UK (see Coats & Max, 2005).

Tables 3 and 4 (see following page) shows the large gaps that can exist between a manager's perceptions about a topic and the empirical reality represented by the benchmarks. Two of the factories are quite good on the production topic, with small gaps, but on site ground contamination (to groundwaters), all the factories have dubious knowledge. On other important topics such as »training« and communication, there are very large gaps with significant consequential problems. In the example, it is quite likely that the production and quality managers will agree on things being well under control and this gives a slight bias to any overall conclusions. This type of bias should not exist with the second table.

The illusions also tie in with Morgan's *psychic prison* wherein people exposed to an unreal situation long enough will view it as *real* (as in Plato's cave); hence the psychic prisons of hierarchical constructs could be leading people to reject new knowledge as not conforming to their surrounding norms or context-constructed realities. Another reality might also be super-imposed, e.g. »if I disobey the rules I get sacked and my family

Table 3: Gaps between empirical reality and irrational perceptions (production)

factory	process characteristics affecting run-off and groundwaters question 10 »There is a strict control of production processes«	managers Q10 score	bench mark Q10	gap
1	process chemicals quite well controlled low risk of contamination	5/6/9/10	9	0–4
2	poor control of some oils/chemicals major risks of contamination	3	3	0
3	poor control of some oils/chemicals major risks of contamination	7/10	3	4–7
4	good control of process chemicals and oils low risk of contamination	8/10	8	0–2
5	poor control of toxic chemicals major risks of contamination	8/9	2	6–7

Table 4:   Gaps between empirical reality and irrational perceptions (site)

factory	site characteristics affecting groundwaters	managers	bench mark	gap
	question 13 »groundwaters are no problem under our site«	Q13 score	Q13	
1	major bare permeable soil (possible oil spills), leaking sewage pipes (ammonium in well water)	3/6	6	3–0
2	significant badly contaminated bare soil, damage already done; sewage has occasional highs	5	0	5
3	significant badly contaminated bare soil, damage already done; bad spills still abound	8	0	8
4	mostly concrete, few spills, permeable soil	7/10	5	2–5
5	significant badly contaminated bare soil, damage already done; bad spills still abound	7	0	7

starves!« In this respect there can be links from the psychic prison to personal mental models as self-reinforcing *traps,* made of a reality that can impede (further) learning and encourage passivity.

In summary the literature centres firstly on psychic prisons and mind viruses which can cause different realities in one place, e.g. between the parachutists, training managers and the others in production. Secondly, there is a focus on the sociological relationships of perceptions, knowledge, decisions and hierarchy where, in the case study factories, there is a difference in culture and behaviour between top, middle and lower management. All these are acted out in a context of an authoritarian *aggression* from top management, and the (deluding) passivity of people in the lower levels of management who construct a *space* for themselves by purposeful modifications of instructions – the elision and conflation of Archer (1996) and the cheats of Mars (1994).

THEORETICAL AND PRACTICAL IMPLICATIONS

There is an academic view (Archer 1996) of a logically coherent top management culture undergoing no dilution or change as it passes through various management layers to the operational level. In the case of the study factories this logical coherence does not exist and there is mutation of meaning at every level of socio-cultural and hierarchical interface and this itself varies with the context of the topic. Within the Polish survey data there were very few core groups of like-perception socio-cultural groupings in a factory, with sometimes only a couple of groupings as with training and communication. Mostly, the perceptions are spread across the *strength-of-feeling* score range (0–10) as with innovation and environmental responsibilities.

This fragmentation of score levels suggests that there is no common cognitive or perceptive view or knowledge to provide a base from which good decisions may be made (Frost et al., 1989; Zey, 1992). This in turn suggests that there was no coherent mission in the factories and giving a resultant general passivity in relation to making beneficial changes. The parachutists sent in to change the culture admitted to having little effect.

Figure 2: Some of the main findings from the research

On average, the managers in the first survey acknowledged that 46% of the decisions they made were to selfbenefit
and, for the second survey there:

- were strong indications of dysfunctional hierarchies which allow communication elision and some (managerial) agency domination leading to ambiguity in company direction;

- was a failure of meanings, knowledge and other communications through the management level interfaces leading to a wide gap between empirical reality and manager's perceptions, hence a lack of appropriate decisions and actions this being possibly due to:
- a lack of a logically coherent, top management culture representing some company vision or mission, the absence of which was giving fragmented perceptions and deviant socio-cultural groupings.

Particular problems in the dissemination of knowledge from the environmental experts who are apparently hoarding this as an instrument of power (in order to keep their jobs), thereby reducing the opportunities for management and environmental improvement.

DISCUSSION

The Passive:Aggressive Organization and Socio-cultural Patterns

The research established the presence of quite serious problems in the factories (Figure 2). The companies had dramatically different groupings of perceptions, but all exhibited the characteristics identified by Neilson et al. (2005). From these findings it was possible to induce conclusions about how and why knowledge gaps relating to the technical aspects of managers' work were formed, and then how, and if, these gaps were influencing the types of decision that were being made, and the underlying endemic passivity behind this.

In the literature there is extensive coverage of hierarchies and organization (e.g. Weber, 1922; Belbin, 2000; Drucker, 1993; Coulson-Thomas, 1997; Peters, 1982, 1987) but, in practical terms within the factories, and with all the nuances of interpretation, a problem remains of either how to reconcile an individual's *free-will* (Badcock, 1991) within an organization, or how to control it whether under democratic or communist contexts. In the study factories there is the titular control of the hierarchy, but most of the managerial allegiances lie elsewhere, either in informal sub-cultures and socio-cultural groupings within the factory (different in each factory) or in relationships that lie outside in the community.

Discussion

In defining and describing Weberian bureaucratic cultures, still alive and well in post-socialist Europe and pseudo-capitalist China, further research must determine if it is the hierarchical nature of civil and public organizations which determine the organizational culture or whether it is the mind-set, e.g. Confucian philosophy or Weberian precepts, of budding bureaucrats that facilitate such cultures in complete disregard of the national politics or factory needs.

In the context of the Neilson proposals, are top managers inherently, perhaps genetically as well, aggressive in the workplace and, are hierarchical organizations characterized by aggression at the top and passivity lower down the management chain, and how would this fit with the Neilson et al. prognosis?

In their proposals for remediation strategies they stress five essentials:

- Getting the attention of top management;
- New Blood to facilitate change;
- Leave no building block unturned, change everything quickly;
- Make decisions and make them stick;
- Spread the word and the data.

In the organizations researched as above, the top management did realize the need for change and the breaking down of local fiefdoms, it also introduced *new blood*, the parachutists, and everything was changed in the five factories very quickly (overnight in some cases), and the decisions were made to stick, and stuck, in a sea of non-cooperation. So what went wrong?

Two authors (Ramsey, 1996; Coulson-Thomas, 1997) mention the failure rate of management fads and quote figures of 70% to 80%; the research quoted in this paper outlines the types of things that go wrong, and why. The data indicated that any mission or vision created by top management was not adopted in the heart by the lower levels of management because it was poorly communicated or misconstrued, deliberately or otherwise. The parachutists who were brought in to *re-culturize* the management strata failed to realize or acknowledge that there were existing strong socio-cultural groups exhibiting strong entrenched passivity, which were unwilling to be coerced into new ways of thinking.

The »word and the data« may have been spread by dictat but they were re-interpreted at each level of management. The situation was exacerbated in that each decision taken by top management was presented as an instantaneous fait accompli, hence there was no opportunity for any synthesis in modifying dubious decisions (Frost et al., 1991). An example of this was the sacking of mature operators in favour of taking on untrained people at a much lower salary level, the result was a significant loss of expertise and a lowering of product quality together with an emotional retrenchment by middle managers who were *standing by waiting for things to happen* (or to »not happen«).

Remediation strategies

Cloke & Goldsmith (2002) quote Lewin in his writings on group dynamics which, from surveys, showed that it is possible to move away from Taylor-like time and motion and bureaucratic controls to one of humans-oriented structures which encouraged and developed group interactions, communication and motivation. If this had have been in place in the factories above, then it is extremely unlikely that the middle managers would have been *standing by*. The went on to note that in most organizations there is close attention to achieving goals, objectives and targets, with less attention being paid to the values implications and ethical significance of these goals and the means used to achieve them.

>*When organizations ignore values, ethics and integrity and operate on
the basis of laws, rule sets and regulation, they reduce their values to con-
formity and a narrow self-interest of not getting caught.*«
(Lewin, 1947)

The implications of the Neilson et al. precepts shown in the introduction imply that, if the corollary of those points were implemented, then
there would be a good organization. Moreover, their recipe for success in
changing the organization implies rule sets and top-down communication
with no consideration given to ethics and culture.

>*Organizations can impose rules through more or less coercive methods
allowing employees to bypass personal responsibilities, or they can encourage employees to develop their own values and ethics in dialogue
with each other.*«
(Lewin, 1947)

Before doing anything else, it is necessary to determine, by the use of outside resources, what the exact socio-cultural characteristics of the organization are; on past experiences detailed in the literature, without this
knowledge, whatever is done to change the organizational culture will be
fruitless. And the results must inform the following strategies:

The strategies for altering the various mind-sets throughout management
should include the formulation of a logically coherent culture (Archer,
op cit) which could need to be communicated throughout the organization; this presupposes that the language will be unambiguous throughout (Holmes, op. cit.). This culture should include a company vision or
mission which is intelligible, meaningful and acceptable to the whole organization.

In order to remediate the passiveness of the managers, it is proposed that
top management introduce a new way of looking at efficacious organization which includes a concentration on ethics and values in the work-

place (Cloke & Goldsmith, 2002) – this would also be included in the company mission and vision.

Part of the ethics and values must be the transmutation of the organization from bureaucracy and autocracy to democracy and collaborative self-management (Cloke & Goldsmith, op. cit.) which can be part of a business process re-engineering (Coulson-Thomas, op. cit.) incorporating »webs of association«. The latter implies that the managers should function as teams and with unambiguously empowered authorities in decision making.

To ensure that manager expectations can be satisfied at the human level, Vroom's Expectancy Theory (1964) could be used to facilitate employee motivation and emphasize self-interest in the alignment of rewards with employee's wants. In the connections between expected behaviors, rewards and organizational goals, the theory applies, and is substantiated, even if the emplyer has not committed to a stated or unstated manager expectation (see also Struzyna & Dyduch, 2003).

CONCLUSIONS

In the comparison with the data from this survey and the Neilson et al. framework, it would seem that the Passive:Aggressive organization is a generic model which identifies the problems and solutions from a top level management viewpoint, but neglects existing socio-cultural impediments to change and the remediation of aberrant socio-cultural groups at lower levels. The sacking of all the managers in these groups could be a disaster to the organization.

It is suggested that the remediation strategies outline above would enable the lower level managers *buy in to change* before the formal introduction of new management systems and would also remove the top manager to subordinate manager disconnect.

It is concluded that Passive:Aggressive model does, at a surface level, characterize the five Polish companies surveyed in the initial researches but, like other models or fads, has little chance of empirical success without attention the human socio-cultural aspects.

REFERENCES

Alasuutari, P. (1995),
Researching Culture – Qualitative Method, Sage, London.

Archer, M. (1996),
Culture and Agency – Culture in Social Theory, Cambridge UP.

Badcock, C. (1991),
Evolution, co-operation and human nature – the issue of free will,
Basil Blackwell Ltd, Oxford.

Belbin, R (2000),
Beyond the Team, Butterworth-Heinemann, Oxford, UK.

Bergami, M. and Bagozzi, R. P. (2000),
Aspects of Social Identity in the Organization,
British Journal of Social Psychology Volume39, December 2000.

Cameron, K. S. and Quinn, R. E. (1999),
Diagnosing and changing organizational culture,
Addison Wesley, Longman, Harlow, UK.

Clegg, S. R. and Palmer, G. (1996),
The Politics of Knowledge Management, Sage, London.

Cloke, K. & Goldsmith, J. (2002),
The end of management, Jossey-Bass, San Francisco, USA.

Coats, D. & Max, C. (2005),
The Work Foundation, The London Health Commission.

Coulson-Thomas, C. (1997),
The Future of the Organiation, Kogan Page, London.

Craig, J. H. S. (2004),
Theorizing Post-socialist Management from Action Research in Poland,
presented at the Employment Research Unit Conference at Cardiff
University in 2004.

Craig, J. H. S. & Lemon, M. (2005),
*Perceptions and reality in the implementation of formal management
systems: a study of the human interface in Chinese and Polish factories*,
presented at the IET Humans and Systems Conference, London 2005.

Craig, J. H. S. & Lemon, M. (May 2008),
*Perceptions and reality in quality and environmental management
systems: a research survey in China and Poland*,
The TQM Journal, Vol. 20 No. 3, Emerald Group Publishing.

Drucker, P. (1992),
Managing for the Future, BCA/Butterworth-Heinemann, Oxford.

Drucker, P. (1993),
Post-capitalist Society, Harper Collins, New York.

Fritz, R. (1999),
The path of least resistance for managers, Berrett-Koehler.

Forsyth, P. (2000),
How to Motivate People, Kogan Page, London, ISBN 0 7494 32551.

Frost et al. (1989),
Reframing organizational culture, Sage, London, ISBN 0 8039 3650 8c.

Goodenough, O. R. (1995),
Mind Viruses (& culture), Social Science Information,
Vol. 34 No. 2, Hegel (in Morrrison) 1995.

Heugens, P. (April 2005),
A Neo-Weberian Theory of the Firm, Journal of Organization Studies,
Vol. 26 (4), 547 – 567, Sage, London.

Holmes, J. (2001),
An introduction to socio-linguistics,
Pearson Education, 2nd edition, Harlow.

Lewin, K. (1947),
*Frontiers in group dynamics – Concept, Method and Reality in Social
Science, Human relations*, hum.sagepub.com Likert Scale (1932).
http://en.wikipedia.org/wiki/Likert_scale

Merton, R. K. (1967),
Social Theory and Social Structure,
Collier-MacMillan, London.

Morgan, G. (1986),
Creative Organization Theory,
Sage Publications, London.

Morgan, G. (1998),
Images of Organization, Sage, USA.

Morrison, K. (1995),
Formations of Modern Social Thought (Marx, Durkheim, Weber),
Sage, London.

Neilson, G. L., Pasternack, B. A., Van Nuys, K. E., (2005),
The Passive:Aggressive Organization,
Harvard Review, October 2005, 83 – 92.

Patten, R. et al. (1996),
The new management reader,
Routledge, London.

Perrow, C. (1972),
Complex organizations,
Scott Foresman, Glenview, USA.

Perugin, M. and Bagozzi, R. P. (March 2001),
Theory of Planned Behaviour,
British Journal of Social Psychology, Vol. 40 pt1.

Ramsay, H. (1996),
Managing Sceptically – A Critique of Organizational Fashion,
in: The politics of knowledge management, edited by Clegg S.R.
and Palmer G., Sage, London,155–172.

Skrupinska, K. (2004),
Why Worker Participation,
Human Resource Management,
Institute of Labor and social studies, Polish Academy of Sciences.

Sperber, D. (1996),
Explaining Culture – A Naturalistic Approach,
Blackwell, Oxford.

Struzyna, J. and Dyduch, W. (2003),
A method for building company capital,
Human Resource Management, Institute of Labor and social studies,
Polish Academy of Sciences.

Thomas, R. and Davies A., (2005),
*Theorizing the Micro-politics of Resistance: New Public Management
and Managerial Identities in the UK Public Service,*
Journal of Organization Studies, Vol. 26, May 2005, 683–706, London.

Townsend, R. (1970),
Up the Organization, Michael Joseph.

Vroom, V. H. (1959),
Some Personality Determinants of the effects of participation,
Journal of Abnormal & Social Psychology, 59, 322–327.

Weber, M. (1922),
Wirtschaft und Gesellschaft,
Tübingen.

Zey, M. (1992),
Decision Making – Alternatives to Rational Choice,
Sage, London.

RETHINKING SUSTAINABILITY IN ACADEMIC EDUCATION

Katja Kuhn

SRH University Heidelberg, School of Engineering and Architecture, Dean of Studies

»I can't conceal the fact that I've become a bit nervous about whether we'll be able to do it,« German Chancellor Angela Merkel said on a press conference after the summit in Copenhagen. »We all know time is running out and we need to get serious.«

This is not a new statement. In 2000 United Nations former Secretary General Kofi Annan characterized the three central challenges facing the international community at the beginning of the new millennium as helping the peoples of the world to achieve »freedom from want, freedom from fear, and the freedom of future generations to sustain their lives on this planet«.

Principles of environmental education include the fundamental elements of sustainable development: the need to consider social aspects in environmental education and take into account the close links between economy, environment and development; the adoption of both local and global perspectives; the promotion of international responsibility and the involvement of as many actors a possible.

During the last decades various conferences and organizations have offered definitions of environmental education. In this context, the terms education for sustainability and environmental education are used as an interdisciplinary umbrella approaches where environmental education includes the economic, environmental, and social dimensions contained in the concept of academic education for sustainability. As such, it embraces components from traditional disciplines such as social sciences, engineering, political science, geography, as well as others.

The question of sustainable approaches in education is no longer a question of transmitting exogenous knowledge: »education starts from the community experiences and search for possible solutions to significant problems« (Maya, 1993). Maya (1993) argues that an environmental education is related to the construction of a participatory society. Education based on participatory investigation brings scientific information as well as methodological tools to societies who will be able to construct their own development (p. 12). The »Treaty of the Earth Council« in 1993 put forward the expression of an environmental education for sustainable societies and global responsibility. The main topics mentioned in this content are solidarity, freedom from alienation, profound social transformation, questioning of the dominant socio-economic systems and the current growth-based development models. Many paradigmatic characteristics (the symbiosynergic socio-cultural paradigm and the corresponding inventive educational paradigm) as identified by Bertrand and Valois (1992) can be found here. According to them, education is not directly associated with the transmission of predetermined knowledge, but with the production of new knowledge in a co-operative and critical process instead. Sustainable education therefore is linked to the conception of sustainability-as-a-joint-societal-project. Emphasis has to be placed on an educational perspective (education for personal and social development in relation to sustainable development approaches) rather than on exclusive environmental preoccupations. Sustainable Development is not perceived from a resource management perspective according to a rational and exogenous economic logic. Instead, it is defined as »the community's competency to interpret its own problems, natural resources, needs and aspirations and to develop creative projects that minimize social and environmental costs« (Ossa, 1989).

In a sustainable world, environmental protection as well as economic objectives and social responsibility have to be linked. Many educators support societies to achieve sustainability by teaching the three »E's«-environment, economics, and equity along with the traditional subjects like mathematics, physics etc. They are fostering awareness of sustainability among students, communities, institutions, and governments but often they do not integrate interdisciplinary ways of thinking and understanding of sustainable education. In coming decades, education for sustainability has to have the potential to serve as a tool for building stronger bridges between the classroom and enterprises, and between academic institutions and societies.

As attention to the concept of sustainability can be found domestically and abroad, efforts must continue to bring all stakeholders together in its pursuit. Societies, enterprises and governments should consider increasing their efforts to ensure that thoughtful, comprehensive planning is promoted by the formal and nonformal education community. This means that we have to follow a constructivist approach in order to implement sustainable ways of learning and thinking and to develop a holistic view on environmental issues and sustainable understanding.

Efforts should focus attention on the delivery systems used to achieve these goals. Key questions could be »do we educate young potentials respectively a group of technical-policy-managerial professionals potentially proficient in guiding national industries, communities, and governments according to sustainable standards?«

Research reveals an important need for professionals with increased knowledge of environmental complexities as well as the integrative skills needed for understanding the interdependent relationships between environment and economy. Education, especially on the academic level, will play the key role in responding to this need.

If a sustainable management is to be achieved, academic institutions are supposed to take over a leadership role, breaking new ground to prepare societies for an age of change in a world of cultural diverse and growing populations, global markets, and changing environments. Hence, education for sustainability requires an understanding of the interdependence and interconnections of humans and their environment. Sustainable concepts have to be set up interdisciplinary and include knowledge of global socio-geopolitical disciplines, technical and physical sciences, and human socio-economic systems. To the extent possible, educational curricula and didactical concepts should reflect the interrelationship among disciplines that are central to sustainable development. The benefit of this approach is twofold: sustainability can be used as a principal theme for encouraging integrative thinking and secondly for activity-based-learning approaches. Learning about sustainability necessitates breaking down the walls between disciplines, perhaps by focusing on one practical example addressed through various perspectives. Ideally, courses with social, economic, or environmental content should be accompanied by interdisciplinary subject matter on sustainability, which draws from a number of content disciplines. Nevertheless processes and resources used to integrate education for sustainability across the curriculum will remain a regional issue assessed by communities and their respective stakeholders. Regionally focused materials and specific, hands-on examples will have to be developed, and students and instructors should benefit from training and practical assistance. In order to support this kind of experience, existing education standards may need to be revisited to embrace the major elements of sustainability and in order to identify variables, leading to crossover effects between the academic world, businesses and society.

Universities have a unique role to play in this emerging environment particularly in the economic growth and sustainable development of the communities in which they exist (Gnuschke, 2001; Hill, 2001).

In contrast to the past, when university scholars were discouraged from becoming entrepreneurs, a new generation of university faculty is now closing the gap between the world of academic scholarship, economic activity and sustainable approaches. Therefore, the evolving university as centre of intellectual property can be seen as an engine for sustainable development, economic growth and prosperity (Gnuschke, 2001).

Accordingly, higher education institutions must act not simply by improving quality through proper implementation of existing standards, but more fundamentally, by reviewing the relevance of the standards that have been used and by developing sustainable interdisciplinary approaches integrated into their academic curricula.

Enterprises require staff that is on the one hand side environmentally literate and on the other hand skilled in interdisciplinary systems approaches to solving problems. In addition, the establishment of global markets requires staff that is familiar with different cultures and habits as well as environmental background information. Possibilities of cooperation with the academic world are therefore numerous: Businesses can support formal education by participating as experts and mentors, by offering internships, by providing employees as well as academic instructors with opportunities for advanced training, and by employing business sites as classrooms. Most importantly, the business community and the educational institutions can engage in ongoing dialogue about future developments, common goals and how best to achieve them. In order to reach these goals, in some cultures the rise of a new understanding between the role of enterprises and the academic world seem to be inevitable, but if a joint agreement about each others role and the use of potential synergies has been reached the expected output can be a very important achievement.

Federal, state, and local governments can support sustainable educational activities in the public and private sectors and build intergovernmental alliances to advance environmental education and training by supporting educational activities. Educational institutions should seek ways to collaborate and develop in this context a kind of entrepreneurial understanding. Additionally it is necessary for governments to support these synergies as they will play the role of policy makers providing the necessary framework to encourage regional sustainable involvement. Tools used could be based on innovative policies or on incentives for the universities and industries to expand their joint research and development activities.

The dynamic relationship between government, industry and universities has been identified by Etzkowitz as »triple helix« (Etzkowitz, 1994). The model engages the university as the centre of excellence with academic-based research and development activities, identifies enterprises as the provider of the customer demand based on its commercial as well as research and development activities, and assigns governments the role as a policy maker. The integration of these different actors lies at the heart of the triple helix system that ideally will create regional or local knowledge spillovers; thus, increasing the competitive advantage of economic development, either regional or national.

In practice the implementation of such strategy requires a greater science and technology policy capacity on the state, industry and academic side, since the judgments of the level and type of intervention and/or cooperation in particular areas become more critical (Etzkowitz, 1997).

The key issues are the synergies among the three different actors – in different societies reflecting different traditions of political economies, and different levels and types of economic development, including the macro and micro economics of each region.

Referring to sustainability and sustainable education we will need to establish comparable structures found and described in the model of the triple helix between enterprises, the academic and political world.

But while working on such concepts we have to take into consideration the role of different cultures und cultural behavior. Different societies in the world will choose different approaches to work on environmental education. In the past, as in Germany, industry often grew out of the links between professors and students – links that were lost as soon as a new generation of managers took over. In Germany today students with practical training and experiences have become the new driving force and academic institutions do appreciate this development. It is good practice at some German universities nowadays to bring in experts from industry to review their programs and then go into the world looking for connections and partners.

Unfortunately there is still a lack of an intellectual concept of how the »crossover« process between the academic world, enterprises and communities works, how it can be initiated effectively, how the structure can be improved and how synergies between three different actors can be stimulated most efficient. There is both a great need and a great deal of enthusiasm for systematically and critically comparing experience with knowledge systems across a wide range of sectors and regions.

The international community has already gained fundamental experiences regarding the application of research, observations, assessment, and consulting or knowledge management systems that have been designed to foster goals of economic prosperity, human development, or environmental conservation – examples include the international communication research system, the world's campaigns against HIV and AIDS, and efforts to reduce global warming and the destruction of the ozone layer. Still many international efforts for sustainable development have been initiated and developed ad hoc, learning and using little from relevant social science knowledge, additional efforts in other scientific areas, and reflection on own experiences. Experiences have rarely been critically examined to determine what lessons they offer contemporary efforts to build more effective decision-making-systems for sustainability. As a result, we know less than we could about the functioning of the system, which kinds of knowledge systems work at all under what conditions. Previous studies (e.g., Ruttan et al., 1994) have identified two general features of systems that are able to link knowledge and action successfully: (1) organizational and institutional linkages between the suppliers of knowledge and their users (i. e., bridging institutions) and (2) recognition that location-specific needs must be taken into account when developing usable knowledge. Although this earlier work has provided important insights into what makes some systems successful in linking knowledge and action, it only considered a few areas of research and did not focus on the barriers that prevent success (Cortese, 2003).

Connecting knowledge from R & D systems with concrete action for sustainable development is a complex process. Efforts to link knowledge with action demand for some research in response to articulated needs of decision makers, rather than only in response to interests of researchers. It has proven difficult to ensure that research informs decisions, even in circumstances where a system is developed explicitly with the goal of affecting decisions, such as some decision-support systems which are often not used by the intended user (Cortese, 2003). Systems that successfully link knowledge with action involve various groups e.g. knowledge producers (e.g., researcher, scientists and engineers); knowledge consumers (decision makers, such as managers or politicians); and program providers who often bridge those two groups, attempting to ensure that what the knowledge producers assist the users in making their decisions and in taking action (Cortese, 2003).

It will be important to develop a system of interdisciplinary educational approaches which will not focus too narrowly on the protection of natural environments (for their ecological, economic or ethical values), without taking into account the needs and rights of humans and their cultural background associated with their environments, as an integral part of the ecosystem. In addition it will be also necessary to emphasize aspects related to contemporary economic realities, by placing greater emphasis on concerns for global responsibility and to involve academic institutions, business as well as governmental agencies in the development of sustainable education concepts.

The idea of environmental education will demand for a modified understanding 1) of the environment, 2) of education, and 3) of sustainable development and 4) of the responsibilities of the relevant actors.

In such a system different conceptions of environment, education and sustainable development may coexist. These concepts influence the way societies and their educating institutions define and practice sustainable education. And it is this diversity and the related approaches that should be valued as »fuel« for critical reflection, discussion, contestation, and exchange.

The complexity of environmental education has to be seen with other interrelated educational dimensions: e.g. intercultural education, international markets or international engineering education and from the perspective of the involved key players: academic institutions, enterprises and government. Joint efforts will have one thing in common: The ultimate goal in education therefore should be the development of responsible societies in sustainable economies where sustainable development is one of the expected outcomes.

REFERENCES

Bertrand, Y. and Valois, P. (1992),
École et sociétés. Montréal: Éditions Agence d'Arc.

Cortese, A. D. (2003),
The Critical Role of Higher Education in Creating a Sustainable Future.

Etzkowitz, H. and Kemelegor, C. (1997),
Research Centres: The Collectivisation of Academic Science,
Minerva, 35:3.

Etzkowitz, H. and Mello, J. M. C. D. (1994),
The Rise of Triple Helix Cluster: Innovation in Brazilian Economic and Social Development, International Journal of Technology and Management & Sustainable Development, 2:3, 159–171.

Gnuschke, J. (2001),
Intellectual Property and Economic Development,
Business Perspective, Summer/Fall.

Hill, H. (2001),
Small and Medium Enterprises,
Indonesia Asian Survey, 41:2, 248–270.

Maya, A. A. (1993),
Perspectivas pedagógicas de la Educación ambiental –
Una visión interdisciplinaria,
in F. Otalora Moreno, E. Y. Castillo and C. L. Higuera (Eds.),
Cuaderno de Trabajo – Serie Estudios Ambientales (pp.), 2,
Florencia: Universidad de la Amazonia.

Ossa, L. A. (1989),
Educación ambiental, Ecologia, 3, 3.

Gnuschke, J. (2001),
Intellectual Property and Economic Development,
Business Perspective, Summer/Fall.

Hill, H. (2001),
Small and Medium Enterprises,
Indonesia Asian Survey, 41:2, 248–270.

SUSTAINABLE USE OF RESOURCES FOR AGRICULTURAL PRODUCTION IN PENINSULAR MALAYSIA: SOIL QUALITY PERSPECTIVES

S. Zauyah, R. Noorhafizah, B. Juliana, R.A.W. Aishah and H.A. Azizul

Department of Land Management, Faculty of Agriculture, Universiti Putra Malaysia, 43400 Serdang, Selangor, Malaysia

INTRODUCTION

The organic carbon, pH and heavy metal contents in agricultural soils are some of the indicators of soil quality and sustainable soil management. In Peninsular Malaysia, organic carbon may be added through mulching of oil palm empty fruit bunches fronds and organic fertilizers such as chicken dung. Due to low pH, liming materials are added before crop production. Heavy metals in the soils may be added through chemical and organic fertilizers and also soil amendments such as lime and waste materials. Our research for the past decade includes assessment of these indicators in agricultural soils planted with different types of crops from several parts of the country. For background levels to compare the values obtained from cultivated areas, the properties of soils from adjacent forests and saprolites were also investigated. Other investigations had included not only the total heavy metal concentrations but available heavy metals, metal fractionation and microbial biomass carbon and nitrogen. The Malaysian soil investigative levels for heavy metals were established by Zarcinas et al. (2004). The 95th percentile of heavy metal contents from 240 agriculture soils were proposed to be the investigative levels. Soils with values over this 95th percentile will be considered to be contaminated. This paper highlights some of the results of the above studies.

TYPES OF SOILS IN PENINSULAR MALAYSIA FOR AGRICULTURAL PRODUCTION

The common soil types used for agricultural production in P. Malaysia are Entisols, Inceptisols, Histosols, Ultisols and Oxisols. Our assessment of soil quality is concentrated in some of these areas, from the western lowlands (Entisols, Histosols Inceptisols, Ultisols) to the hills of Cameron Highlands which is mainly Ultisols. The Entisols also include soils from tin ex-mining areas. Plantation crops such as oil palm are mainly planted on Inceptisols, Histosols, Ultisols and Oxisols, while rubber is mainly on Ultisols and Oxisols. Leafy and fruit vegetables are planted on Inceptisols, Histosols and Ultisols, including the highlands while the root crops are concentrated in the ex-mining land and also the loamy Ultisols. Paddy is planted on Entisols and Inceptisols along the western coast concentrating in Perlis, Kedah and Selangor.

METHODS OF ANALYSIS

The assessment of the agricultural soils include total heavy metal concentrations, available heavy metals, heavy metal fractionation, organic carbon, pH, cation exchange capacity, total nitrogen, microbial biomass carbon and nitrogen and spatial variability of chemical properties and heavy metals.

Total heavy metal concentrations

Total heavy metals were extracted through digestion with aqua regia, containing a mixture of concentrated hydrochloric acid (HCl) and nitric acid (HNO_3) in a ratio of 3:1. The metal concentrations in the extracts were determined by flame atomic spectrometry (AAS)

Available heavy metal concentrations

Available heavy metals were extracted using three different extractants, i.e. ethylenediaminetetracetic (EDTA), diethylenetriaminepentacetic

(DTPA) and 0.1 M hydrochloric acid (HCl) as outlined by Van Ranst et al. (1999).

Heavy metal fractionation

Total elemental contents provide little information on the mobility and bioavailability of heavy metals. The mobility and bioavailability of heavy metals depend on their physical and chemical forms. Tessier et al. (1979), developed an extraction scheme, which allows the division of the total metal content into five fractions (exchangeable, carbonates, Fe/Mn oxides, organic matter and residual). Heavy metals in available fractions such as exchangeable and carbonate fractions may indicate the potential forms of metals that can be accumulated in the plant. The sequential extraction procedures used in our study are modified from the schemes developed by Tessier et al. (1979) and Shuman (1979). The extraction was modified for our soils because these soils do not contain carbonates and appreciable sulphides and manganese. The first two steps (for exchangeable and organic bound fractions) are those from Tessier's scheme while the third step (for amorphous iron oxides) followed that of Shuman (1979). Step 4 is the determination of heavy metals in the residual fraction using aqua regia instead of nitric, hydrofluoric and perchloric acid. All the analyses were performed in duplicates. Extractions were carried out on 1.0 g of soil and involved the following steps:

F1: Eight ml of 1 M $MgCl_2$ were added to the sample and suspension was shaken for 1 h and then centrifuged (20 min, 4,000 rpm).
F2: Six ml of 0.02 M HNO_3 and 10 ml of 30% H_2O_2 were added to the residue obtained from the first extraction, and the suspension was shaken for 5 h at the temperature of $85 \pm 2°C$. After cooling, 10 ml of 3.2 M CH_3COONH_4 were added and shaken for 30 min and centrifuged.
F3: The samples were extracted with 20 ml of solution 0.2 M ammonium oxalate and 0.2 M oxalic acid, shaken in the dark for 4 h and centrifuged.
F4: The heavy metals contents in the residual fractions were determined by aqua regia method.

Cu, Ni, Zn and Pb were determined by atomic absorption spectroscopy (AAS).

Soil pH, organic carbon, total nitrogen, available P and exchangeable K

Soil pH was determined with a digital pH meter in the supernatant suspension of 1:2.5, soil water ratio. Organic carbon was measured using Walkley & Black method and cation exchange capacity was determined as described by Van Ranst et al. (1999). The total N (TN) of the soil was determined using Kjehdahl Method (Bremner and Mulvaney, 1982). Available P was determined using the Bray and Kurtz II method and exchangeable K was extracted with 1 M NH_4OAc, pH 7.0 using the leaching method and determined using atomic absorption spectrophotometer.

Soil Microbial biomass Carbon and Nitrogen

The microbial biomass has been used as an indicator of the microbial contribution to soil organic matter, plant health and understanding of nutrient flux (Kritine and Sara, 2004).

Our study on the microbial biomass is concentrated in areas cultivated with oil palm and rubber adjacent to the forest. Microbial biomass carbon and microbial biomass nitrogen in the soils (0–10 cm) were measured by chloroform fumigation – extraction method (Anderson and Ingram, 1993).

HEAVY METALS IN SAPROLITES AND SOILS
OF DIFFERENT LAND USE

Heavy metals in saprolites and soils

Soils vary across the landscape and parent materials. A study on the heavy metal concentrations in saprolites (uncultivated areas), forest and soils of different land use showed these variations. Soils (0–20 cm) were sampled using stainless steel augers and total heavy metal concentrations were extracted by aqua regia. Only Cu, Ni, Zn and Pb were determined. Table 1 shows the mean concentration (mg/kg) of cultivated and uncultivated soils. Concentration of zinc is highest in all the soils. Most soils cultivated with vegetables have shown a significant increase in Cu, Pb and Zn when compared to the adjacent saprolite or uncultivated soils. The source of these heavy metals is chicken manure which can contain up to about 500 mg/kg of Zn as compared to other organic fertilizers. The use of zinc-based herbicides on mineral soils also resulted in increasing Zn level in the soils. A study by Khanif et al. (1999) showed that pesticides contain 1.3–25% of Zn. The high pH of the soil contributed to the retention of Zn in soil and high proportion of Zn is bound to organic matter. In general however, the values are below the 95th percentile except for Zn in the riverine alluvium and Cu in the soil over landsite.

Heavy metals in ex-mining land

In Perak, large tracts of land close to urban areas were left after tin mining activities which stopped more than 30 years ago. It was estimated that there is about 200,000 ha of ex-mining land in the country. Due to shortage of land for agricultural production, about 5,000 ha of this barren land has been converted to productive areas where vegetables and fruits are grown since the early 1980s. The limitations of these soils to crop production are low organic matter content, low nutrients and water holding capacities, high leaching rate, low biological and chemical activities and high soil temperatures. One remedy to solve all the above problems is the incorporation of organic matter in these soils. The common organic matter used nowadays is chicken dung (almost up to 40tons/ha) due to its quality and availability. For sustainable vegetable production, inorganic fertilizer (NPK) is also applied as top dressing. Although there have been a lot of studies on these soils, the focus has been on the yield potential of different crops and the rates and types of organic amendments. However, the impact of these cultural practices on the environment and the quality of the crops has not been given emphasis. A common quality indicator for both soils and crops can be the heavy metal contents.

Table 2: Chemical properties and heavy metals of cultivated and uncultivated tin tailings

	pH	CEC cmol/kg	O.C %	Cu mg/kg	Pb mg/kg	Ni mg/kg	Zn mg/kg
Cultivated	6.10a	7.74a	0.67a	4.25a	10.47a	3.05a	16.03a
Uncultivated	4.60b	5.61b	0.30b	1.19b	5.83b	1.01b	8.19b
Investigation levels (Zarcinas et al. 2004)				47.3	65.8	41.3	92

Codes in table: a and b indicate significant difference (P ≤ 0.05) between means of uncultivated and cultivated soils across columns

The texture of the soils range from sandy clay loam to sand. Table 2 gives some of the chemical properties and heavy metals of cultivated and un-cultivated soils in the study area. Although there is an increase in values of pH, CEC, organic carbon and the heavy metals of the cultivated soils, only pH, CEC, Cu, Pb, Ni and Zn have significantly increased after cultivation. However, the heavy metal concentrations are all below the investigation levels.

Heavy metal fractionation in cultivated Ultisols and Inceptisols

Some selected cultivated soil samples (26 of Ultisols and 10 samples of Inceptisols) were analysed for chemical fractionation using a modified Tessier's procedure described above. The Cu, Pb, Zn, and Ni fractions expressed as percentages of the sum of individual chemical fractions are presented in Table 3 and Table 4. In the cultivated Ultisols, Cu and Zn showed the highest concentration in the Fe oxide fraction, while Pb and Ni are highest in the residual fraction. The percentage of Cu fractions follows the order: Fe oxides > residual > organic > exchangeable. For Pb, the percentage of Pb fractions follows the order: residual > amorphous Fe oxides > exchangeable > organic. Percentage of Zn fractions follows the order: Fe oxides > residual > exchangeable > organic. The percentage of Ni fractions follows the order: residual > amorphous Fe oxides > exchangeable > organic.

Table 3:   Mean values of Cu, Pb, Zn, and Ni fractions expressed as percentage
            of sum of fraction (%) for the Cultivated Ultisols

	F1 Exchangeable	F2 Organic	F3 Amorphous Fe Oxides	F4 Residual
Cu	1.3	2.5	57	39.2
Pb	7.9	6.3	36.9	48.9
Zn	2.4	1.3	64.0	32.3
Ni	9.0	1.3	15.2	74.6

Table 4:   Mean values of Cu, Pb, Zn, and Ni fractions expressed as percentage
            of sum of fraction (%) for the Cultivated Inceptisols

	F1 Exchangeable	F2 Organic	F3 Amorphous Fe Oxides	F4 Residual
Cu	4.9	8.8	18.4	67.9
Pb	5.7	2.2	3.3	88.8
Zn	0.8	2.3	4.6	92.3
Ni	1.6	1.9	5.8	90.7

The amounts of non-residual fractions (F1, F2 and F3) represents the amounts of active heavy metals while those of the residual fractions may be considered to be the stable form and thus not available to plants for a reasonable period. In this study, the non-residual fractions of Cu, Pb, Zn and Ni in the Ultisols average 60.8%, 51.1%, 67.7% and 25.4% which suggests that the mobility and bioavailability of the four metals are in the order: Zn > Cu > Pb > Ni.

Heavy metals fractionation using a modified Tessier's sequential extraction procedure showed that all the metals in the Inceptisols are dominantly in the residual phase. The general trend in the Ultisols for Pb and Ni is residual > oxalate > exchangeable > organic. For Zn and Cu, the oxalate extractable phase is highest followed by the residual phase. Zinc and Pb contents in the Ultisols are also positively correlated to pH of the soil.

SPATIAL VARIABILITY STUDIES OF CHEMICAL PROPERTIES AND HEAVY METAL CONCENTRATIONS IN PADDY SOILS

One of our current projects is to evaluate the spatial variability of chemical properties in some cultivated soils. The first project started was for paddy soils. Fertilizer management is a major consideration in paddy production. Inadequate fertilizer application limits crop yield, results in nutrient mining, and causes soil fertility depletion. An excessive or imbalanced application not only wastes a limited resource, but also pollutes the environment. With consideration of both economic optimization and environmental concerns, farmers are forced to face with an ever-increasing demand for effective soil fertility management. An approach towards justifying such concerns is site specific nutrient management – which takes into account spatial variations in nutrient status cutting down the possibility of over or under use of fertilizer. There have been growing interests in the study of spatial variation of soil characteristics using geostatistics since 1970s. Few studies have thoroughly investigated the spatial variability of soil chemical characteristics in agricultural fields in Malaysia. Some results on this first set of studies to evaluate the spatial variability of chemical properties and heavy metals in some paddy soils in Selangor, Kedah and Perlis is given below.

Geo-referenced soil samples were taken after harvest, at an interval of 80–90 m, from 84 points within Parit 9, Selangor in April to June 2007. All the soil samples were taken at depth of 0–20 cm, air-dried, mixed and ground to pass a 2 mm sieve, then stored in plastic containers prior to the analyses of soil pH, organic carbon (OC), total nitrogen (TN), available phosphorus (AP) and exchangeable potassium (EK).

Descriptive statistics and correlation analysis were performed using statistical software to determine the relationship between each characteristic measured. Geostatistical analysis using geostatistical software package (GS+ Gamma Design Software) was carried out to examine the within-field spatial variability using semivariograms. Information generated through semivariograms was used for spatial interpolation of unobserved points by a Kriging procedure (Isaaks and Srivastava, 1989).

Table 5:   Descriptive statistical parameters of soil pH, organic carbon (OC), total nitrogen (TN), available P (AP) and exchangeable K (EK)

Area	Characteristic	Mean	Min	Max	SD	CV (%)
	pH	4.79	4.01	5.46	0.20	4
Parit 9	OC (%)	4.02	3.90	4.15	0.04	1
Sungai	TN (%)	0.31	0.17	0.47	0.05	17
Besar (31.52 ha)	AP (mg/kg)	44.35	10.92	172.90	28.28	64
	Logarithm of AP	3.64	2.39	5.15	0.528	14.5
	EK (cmol+/kg)	0.39	0.10	0.80	0.12	31

Codes in table: SD standard deviation, CV coefficient of variation

Descriptive statistics

Descriptive statistics of the measured soil characteristics for all the areas showed wide variations (Table 5). Available P data for the 9 area was not normally distributed and so was e further analyzed by using their logarithmically transformed values (Table 5). Except for pH and organic carbon, all other variables have CV values greater than 14%, the highest being 64% in the case of available P for Parit 11 area. The mean values of total nitrogen exhibited highest value in Parit 11 area, followed by Sawah Sempadan, Parit 9 and Pendang. Mean available P was lowest (5.09 mg/kg) in Pendang.

Table 6:   Statistical values of heavy metals in paddy soils in Selangor

Trace Elements	Sample Size	Mean	Min	Max	SD	CV (%)
	pH	4.79	4.01	5.46	0.20	4
Parit 9	OC (%)	4.02	3.90	4.15	0.04	1
Ni (mg/kg)	84	15.0	4.7	25.7	4.7	31
Cu (mg/kg)	84	16.7	11.0	25.7	3.0	18
Pb (mg/kg)	84	38.0	25.0	59.0	6.9	18
Zn (mg/kg)	84	41.0	25.0	81.0	10.0	24

Codes in table: SD standard deviation, CV coefficient of variation

Table 7:   Best-fitted semivariogram models for soil trace elements and their parameters

Trace Elements	Model	Nugget C_0	Sill C_0+C	Proportion $C_0/(C_0+C)$	Range (m)	R^2
Ni (mg/kg)	Linear	22.12	22.12	0.00	760	0.015
Cu (mg/kg)	Exponential	7.63	15.27	0.501	2110	0.013
Pb (mg/kg)	Exponential	40.9	81.81	0.501	2110	0.110
Zn (mg/kg)	Exponential	57.7	266.2	0.783	1700	0.769

Codes in table: SD standard deviation, CV coefficient of variation

The results of descriptive statistics for heavy metal concentrations are shown in Table 6. The mean trace elements concentration for the area is as follows: Ni (15.0 mg/kg), Cu (16.0 mg/kg), Pb (38.0 mg/kg) and Zn (41.0 mg/kg). In comparison to 95th percentile trace elements values for Malaysian topsoil samples (Zarcinas et al., 2004) in Table 2, the mean values of Ni Cu, Pb and Zn are all lower than the contaminated levels.

Figure 1 shows the semivariogram and fitted models for each trace element measured. The attributes of the semivariograms are summarized in Table 7. The best fit semivariogram model for Ni is linear and exponential for Cu, Pb, and Zn. For linear and exponential models, semi-variance increases with distance between samples (lag distance) to a constant value (sill or total semi-variance) at a given separation distance (range of influence). The ranges for spatial dependence for the trace elements were 760 for Ni, 2,110 m for Cu and Pb and 1,700 m for Zn. The nugget effects (C0), which is the variance occurring at zero distance, were found to be 22.12 mg/kg, 7.63 mg/kg, 40.9 mg/kg and 57.7 mg/kg for Ni, Cu, Pb and Zn, respectively.

The above results showed an example of a study carried out for paddy soils in Selangor. Two other areas are located in Selangor, one in Perlis and in Kedah. All the studies showed that all the heavy metals measured are still below contaminated levels. However, monitoring on soils heavy metal concentration should be done continuously to make sure that the levels remain low. Furthermore, the soil sampling density could be reduced by increasing the sampling interval to reduce the cost and also save time.

Figure 1: Experimental semivariograms of trace elements with fitted models

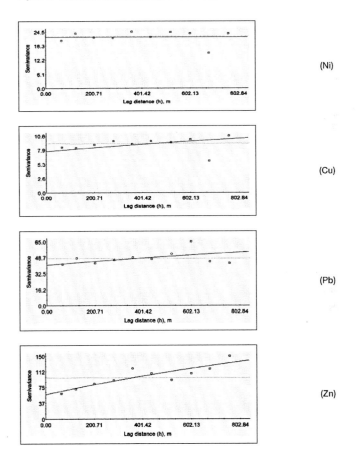

Soil microbial biomass carbon and nitrogen

Another soil quality indicator which is currently being investigated is soil microbial biomass carbon and nitrogen. Soil samples collected from oil palm, rubber and forest area which is adjacent to the plantation in the Bi-

Table 8: Soil chemicals properties and soil microbial biomass. in Bilut, Pahang

Area	pH	T.N. (%)	O. C. (%)	MBC (ug/g)	MBN (ug/g)
Forest	4.5 a	0.48a	3.87a	978a	219a
Oil palm	4.6a	0.32b	3.56a	552b	78b
Rubber	4.88a	0.27b	2.14b	498b	65b

Codes in table: *Means with the same letter are not significantly different at $p \leq 0.05$

lut Valley, Pahang were analysed for total nitrogen (TN), organic carbon (OC), soil microbial biomass carbon (MBC) and nitrogen (MBN).

Table 8 shows that total nitrogen, total organic carbon, microbial biomass carbon and microbial biomass nitrogen are affected by different land use. Forest showed significantly higher TN compared to the oil palm and rubber at $p \leq 0.05$. However, the organic carbon in the rubber area was significantly lower than both oil palm and forest areas. The difference in the microbial biomass carbon and nitrogen showed similar trend to the total nitrogen. Land clearing from forest to oil palm and rubber had resulted in changes of soil properties. More studies have to be carried out on the biochemical properties of the soils to understand the relationship between these soil quality indicators.

CONCLUSION

An assessment of heavy metals in agricultural soils showed that Zn has the highest concentration compared to Cu, Ni and Pb. The paddy soils showed the lowest level for all heavy metals. Although most cultivated soils have shown heavy metal values lower than the investigative levels, monitoring of these values should still be carried out to ensure that crops grown on various soil types are safe for consumption. The utilization of chicken dung especially for vegetable, root crops and fruits had significantly increased organic carbon in the soils. However for plantation crops, it was found that organic carbon content is lower than the adjacent forest and that microbial biomass carbon and nitrogen may be some of the soil quality indicators that can be monitored. It is also recommended that the biochemical properties should be studied in order to monitor the sustainability of our soil resources.

ACKNOWLEDGEMENT

We are most grateful to the Ministry of Science, Technology and Innovation, Malaysia which had funded all the above research.

REFERENCES

Bremner, J. M., Mulvaney, C. S. (1982),
Nitrogen – Total, in: Methods of soil analysis (A. L. Page et al., ed.)
Agronomy Monograph 9, Part 2, 2nd ed., 595–624,
American Society of Agronomy, Madison, Wisconsin.

Gamma Design Software (2000),
GS+, Geostatistics for Environmental Sciences,
Plainwell, Michigan, USA.

Isaaks, E. H., Srivastava, R. M. (1989),
An introduction to applied geostatistics,
Oxford University Press, New York, 500.

Anderson, J. M., Ingram, J. S. I., (1993),
Tropical soil biology and fertility: A hand book of methods,
CAB international, Wallingford, UK.

Shuman, L. M. (1979),
Zinc, manganese and copper in soil fractions,
Soil Science 127, 10–17.

Tessier A., Campbell P. and Bisson M. (1979),
Sequential Extraction procedure for the speciation of particulate trace metals, Analytical Chemistry 51, 844–851.

Van Ranst, E., Verloo, M., Demeyer, A. and Pauwels, J. M. (1999),
Analytical methods for soils and plants. Equipment and management of consumables, Univ. Gent pp 243.

Zarcinas, B. A., Ishak, C. F., McLaughlin, M. J. and Cozens, G. (2004),
Assessment of pollution of agricultural soils and crops in southeast Asia by heavy metals,
1. Peninsular Malaysia. Environ Geochem Health, 26, 343–357.

CARBON SEQUESTRATION AND IMPROVING PRODUCTIVITY OF MALAYSIAN HIGHLY WEATHERED ACIDIC

Agriculture Soils Through Soil Amendment
with Charred Rice Husks, a Biological Charcoal

Rosenani Abu Bakar, Ph.D.

Department of Land Management, Faculty of Agriculture, Universiti Putr, Malaysia
Email: *rosenani@agri.upm.edu.my*

INTRODUCTION

Currently, the potential threat of climate change due to greenhouse gas emissions represents one of the main environmental concerns worldwide. Soil organic matter (SOM) is the largest C stock of the continental biosphere with 1,550 Pg and plays a key role in the reduction of GHG emissions derived from agriculture (Lal, 2004). Therefore, the use of organic wastes as soil amendment is a »win-win« strategy, besides the direct reduction of GHG emissions associated with waste disposal it also increases C sequestration in soil in the form of SOM. However, in most tropical environments, sustainable agriculture faces large constraints due rapid mineralization of SOM (Zech et al., 1997). As a consequence, the cation exchange capacity (CEC) of tropical soils, which is often low due to their clay mineralogy, decreases further. Under such circumstances, the efficiency of applied mineral fertilizers is very low when the loss of mobile nutrients such as NO_3 – or K^+ from the topsoil is enhanced by high rainfall (Cahn et al., 1993). Application of charred biomass or biological charcoal (also known as biochar) as a soil amendment seems to be a promising option to maintain a maximum level of C in soils as charring significantly increases the stability of C against microbial decay (Baldock and Smernik, 2002). Higher charring temperatures also improved exchange properties and surface area of the charcoal. Higher nutrient retention and nutrient availability have been reported after charcoal additions to soil, related to higher exchange capacity, surface area and direct nutrient additions. Biochar can be produced from urban, agricultural or forestry biomass residues, such as, corn stover, rice husks, peanut hulls and wood chips. Biochar from agricultural wastes when used as an amendment has been reported to sequester carbon into stable soil carbon pools, reduce emissions of N_2O, increase crop yields and productivity, aids in soil retention of fertilizer derived nutrients for crop utilization and reduce leaching and run-off to ground and surface waters. Lehmann et al. (2006) calculated and reported that the current potential for rice husks as a biochar is 0.038 PgCyr[-1].

In Malaysia, charred rice husk (CRH) is available at rice mills as a by-product. Rice husk is burnt under reduced oxygen to produce heat for drying rice at the mills, hence the production of CRH which is not fully utilized. This CRH may be a good biochar that can be utilized as a soil amendment, not only to improve soil productivity in crop production and increase C sequestration in acidic and infertile Malaysian soils but also as a waste recycling management option. Currently, very little work has been carried out to investigate the characteristics and potentials of the local CRH as a biochar and soil amendment.

Research objectives

(1)   To determine the physical and chemical characteristics of charred rice husks (CRH) produced at rice mills (total surface area, water retention capacity, negative charge and cation exchange capacity).
(2)   To investigate the effects of addition of CRH as a soil amendment on the water holding capacity and nutrient sorption of selected highly weathered acidic soils, and leaching of plant nutrients.

(3) To determine the effects of application of CRH as a soil
     amendment in a highly weathered acidic soil on the growth
     performance of sweet corn (test crop) and nutrient uptake, soil
     organic carbon (C sequestration) and physical and chemical
     properties of the soil.

PROPOSED PROJECT

Research Methodology

This project would involve laboratory and field studies.

I.   *Collection and characterization charred rice husks (CRH)*
     Charred or burnt rice husks would be collected from a few rice mills
     and analyzed for physical and chemical characteristics (surface area,
     water retention capacity, composition and carbon functional groups,
     negative charge, sorption properties and cation exchange capacity
     (CEC) and effective CEC. Several CRH samples would be taken to
     determine variation in the properties.

II. *Laboratory Investigations: Effects of CRH on water retention
     capacity, sorption and leaching of plant nutrients in highly
     weathered acidic soils*
     Experiments will be carried in the laboratory under controlled con-
     ditions to investigate the effects of different application rates of CRH
     on water retention, pH and sorption and leaching of macronutrients
     (N, P, K, Ca and Mg) in selected soil types with different texture.
     Short term experiments will be carried out using the leaching col-
     umn incubation method. The CRH will be mixed at various rates
     with sieved air-dried soil and placed in a column. Nutrient solu-
     tion would be added at regular intervals and leachate would be col-
     lected to determine volume of solution eluted and amounts of nu-
     trients leached. For this study, samples of several soil types will be
     collected and characterized.

III. *Field experiment: Effects of application of CRH as a soil amendment on nutrient uptake and growth performance of sweet corn, soil organic carbon (C sequestration) and physical and chemical properties.*

A minilysimeter experiment will be conducted under field conditions to determine the effects of CRH application to an Ultisol on leaching and plant uptake of nutrients. Treatments will include three rates of CRH and, with and without recommended NPK fertilizer and with and without plants, with 4 replications and laid out in a completely randomized design. The CRH would be applied and mixed into the top 10 cm soil layer 1 week before sowing sweet corn (test crop) seeds. The chemical fertilizers will be applied in 2 split applications, basal and at flowering. During the experiment, leachate will be collected at regular time intervals for determination of leachate volume and amount of nutrients leached. Soil samples will be collected for characterization before application of CRH and after harvest for determination of total organic C (SOC) and biochemical fractions, total and mineral N mineral N, pH, CEC and exchangeable bases and base saturation. The study will continue up to 3 crop cycles.

A field experiment will also be established to investigate long term field application of CRH on yield of sweet corn, nutrient uptake, accumulation of SOC (C sequestration) and changes in soil physico-chemical properties (water holding capacity, bulk density, pH, CEC and base saturation, total and mineral nitrogen, microbial activities). Treatments will include 4 rates of CRH application (0, 5, 10 and 15 ton/ha) and five replications, laid out in a randomized complete block design; experimental plots of 4 m × 6 m. The CRH will be applied once 2 weeks before sowing the seeds. NPK fertilizers would be applied at the recommended rates (2 split application) in all the plots. The experiment will continue for 3 crop cycles. Soils samples will be collected before treatment application for initial characterization and after each harvest to determine changes in sequestered C, i.e. SOC (total and fractional) and physical and chemical characteristics and microbial biomass. Crop yield and dry matter weight will be recorded and plant tissue analyzed for macro and micronutrient contents.

POTENTIAL APPLICATIONS

Potential use of data and information produced in this study for enterpreuners applying for Clean up Development Mechanism (CDM) projects to produce biochar, not only from rice husks but also other organic wastes as an environmentally safe method of waste disposal or recycling, and for farmers to utilize biochar as a soil amendment to increase soil productivity and crop production. At the same time, contributing to global C sequestration in soil.

REFERENCES

Baldock and Smernik, R. J. (2002),
Chemical composition and bioavailabilty of thermally altered Pinus resinosa (Red pine) wood,
Org. Geochem. 33, 1093–1109.

Cahn, M. D., Bouldin, D. R., Cravo, M. S., Bowen, W. T. (1993),
Cation and nitrate leaching in an oxisol of the Brazilian Amazon,
Agron J. 85, 334–340.

Lal, R. (2004),
Soil Carbon Sequestration Impacts on Global Climate Change and Food Security,
Science 304, No. 5677 (June 11) 1623–1627.

Lehmann, J., Gaunt, J., Rondon, M. (2006),
Bio-char sequestration in terrestrial ecosystems – a review,
Mitigation and Adaptation Strategies for Global Change, 11, 40–427.

Zech, W., Senesi, N., Guggenberger, G., Kaiser, K.,
Lehmann J., Miano T. M., Miltner A., Schroth G. (1997),
*Factors controlling humification and mineralization
of soil organic matter in the tropics,*
Geoderma 79, 117–161.

UV-TECHNOLOGY FOR ADVANCED OXIDATION PROCESSES IN WATER & GAS PHASES

Frank Seitz

IBL Umwelt- und Biotechnik GmbH, Heidelberg, Germany
E-Mail: *f.seitz@ibl-umweltfactory.de*

ABSTRACT

IBL Umwelt- und Biotechnik GmbH presents the *uviblox*® system solutions for the treatment of water and gas phase by UV-radiation providing for a wide range of applications. *uviblox*® system solutions stands for individually designed application oriented systems. We are committed to find the best solution for your problem – always considering the minimized use of resources like energy and other means of production.

Two decades of experience in the field of remediation, process engineering and plant engineering and construction empower us to find the most suitable solution for nearly all kinds of requests not regarding whether it refers to water or gas phase, pilot testing scale or industrial mass production. IBL offers comprehensive consulting services concerning the dimensioning of a UV-unit, process planning, controlling and monitoring as well as complete process development.

INTRODUCTION – WASTE WATER TREATMENT

Needs of development in the field of wastewater and waste air treatment for remediation and industry processes in recent years were caused by rigorous limits (like EU-VOC directive) more and more. On the other hand the traditional treatment techniques like thermic oxidation are very expensive. This brought much attention and investment in the area of Advanced Oxidation Processes (AOP). Among the AOP technologies the UV-photooxidation processes has continuously developed during the past years because of its special characteristics (energy supply, lamp efficiency, reactor geometry etc.). The UV-technology in general isn't new at all in the field of desinfection of drinking water. But the degradation of contaminents with UV-irradiation has never been a market for mass products, cause every case of waste gas and air is different and needs intelligent adapted techniques. Furthermore we passed a time with many developments in this area in the last 5 – 10 years and the developments are still going on. So that we got a technique today – the *uviblox®* technology (*UV-IBL-oxidation*) which suits to a wide range of application for water and gas purification treatment with respect to economics.

The UV-photooxidation technology generally follows the same main rules in oxidation processes by producing highly reactive radicals for oxidizing

Figure 1: Typical *uviblox®* system in operation.

pollutants. They can be produced with H_2O, H_2O2, O_2 activated by special VUV light. Furthermore VUV light can charge organic compounds directly by photolysis. Many organic and oxidizable inorganic substances are the main targets for oxidizing and further destroying in treatment, purification and disinfection of contaminated water, wastewater, air, wastegas and odour by this developed UV technology. Today new efficient systems for the treatment of water and air are available, based on modified Vacuum-UV-medium pressure lamps. The *uviblox*® technology as an AOP technology for degradation and respectively eliminiation of organic substances as well as numerous oxidizable inorganics has reached a much higher degree of efficiency than similar ones.

BASICS

UV-light is a radiation, its energy content is directly depending on its wavelength. So the UV radiation is the most powerful radiation we can use for the technical application of degradation of contaminents. The UV-light can be devided into UV-A, UV-B, UV-C and VUV. VUV (vacuum UV) is the radiation <200 nm and therefore the UV-light with the highest energy density. This radiation can be measured properly only in a vacuum space system, cause this radiation would be absorbed by air. A typical UV lamp can emit different wavelenths. If you want to crack a contaminent directly by UV-light this contaminent has to absorb the emitted energy from the lamp, that means the maximum of emission should be near by the maximum of absorption. This process is called fotolysis. For nearly all substances we know these maxima. They are depending on the types of chemical bonds.

This process of direct photwolysis is also used for the disinfection of water or air. In this case cells, proteins, nucleo acids got there maximum of absorption near 254 nm, the main emitting line of so called low pressure (LP) UV-lamps. So these substances can be cracked, the microorganismen will be inactivated and can't grow or reproduce themselves anymore.

But in the case of VUV the direct fotolysis is not the only effect for degradation. Water and peroxide will be homolized and will generate OH-radicals. Oxygen will be ozonolyzed and the ozone will generate ozone radicals. Both kinds of radicals will cause chain reactions on the organic contaminents. There are further possibilities of combining these processes with fotocatalysis and integration of catalysts online or in series.

Fotolysis

Direct fotolysis of Organics: $\qquad$ $R-R \rightarrow R^\circ + R^\circ$

Ozonolysis

Activation of oxygen (air): $\qquad$ $O_2 \rightarrow O_3$
Fotolysis/homolysis of ozone: $\qquad$ $O_3 \rightarrow O_o$

Homolysis

Fotolysis/homolysis of water: $\qquad$ $H_2O \rightarrow OH^\circ$
Fotolysis/homolysis of peroxide (water): $\qquad$ $H_2O_2 \rightarrow OH^\circ$

TECHNIQUE

For the technical approach you need UV lamps with power suppliers and reactors. A typical UV-lamp is constructed as a glass tube closed on both ends, filled with a mixture of noble gases under a certain pressure, doted with mercury and other metals. By connecting a certain voltage with special frequency a plasma between 2 electrodes will be fired. This plasma emits a spectrum of different UV wavelenths. We differ two main types of lamps, low pressure lamps (LP) and middle pressure lamps (MP), depending on the filling pressure. The low pressure lamps (LP) got a filling pressure < 1 bar and as result two emission lines at 185 nm and 254 nm. This lamps are used for water disinfection all over the world and are mass products. You can realize lamps with 10 to 400 W.

The filling pressure of middle or high pressure lamps (MP) varies from 1 to 10 bar. You can realize lamps with 1,000 to 32,000 W with nearly continous emission spectra. Because you have to find an optimum for each case these lamps are no mass products. The advantages are high power density, middle to high return of VUV, long life time, polychromatic emission, bowlenth to 2 m, few loss of energy at the power supply and small runnig costs.

In former days the supply of the lamps with electrical power was realized by conventional magnetic power suppliers. They were big and heavy with a high energy loss. The UV-emission and lamp power were waving and toddling. They effected high peaks of energy demand for starting and therefore low life times of the lamps. Today we are using special adapted electronic power suppliers (EVGs), which are small, light and gentling the lamps by stepless controlling. You can operate with normal or high frequency.

By collecting lamps in rows and adding lamp row moduls one after the other all sizes of reactors can be realized. Very important is the optimal design of the flowdynamics. In the end the main question is to find with what combination of lamp, EVG, reactor, combined processes we will get the best result for a real special case. The parameters of operation are different for different contents of organics.

APPLICATION

Referring to a long time of experience an steady contact with research and development institutes and universities it is possible to find a solution for nearly every task of purification.

First steps in a project normally is the performing of feasibilty studies to find an answer if the technique can work or not. The next step is the performing of pilot tests onsite for demonstration of the technique and to find the main scale up parameters.

Figure 2:   Performance of tests

lab tests   pilot tests

The *uviblox*® technology can be applied to water phase treatment (WPT) and gas phase treatment (GPT), to disinfection, smell reduction and degradation of organic and anorganic contaminents:

- Disinfection
- Elimination of Odors (Mercaptans, Terpenes, Butter acid, Amines)
- Alcohols, Ketones, Aldehydes …
- Methanes
- Hydrogen Sulfide
- Ammonia
- Exhaust air from stripping plants
- Phenols
- CHC (e.g. vinyl chloride, cis-1,2-dichloroethene)
- Endocrine substances (Antibiotics, Zytostatics)
- COD, TOC
- AHC, PAH, CHC, POL, AOX, PCB
- Cyanides
- Complexing agents
- Endocrine substances (Antibiotics, Zytostatics)
- Pesticides (herbicides, fungicides etc.)
- Organo phosphite (electroplating)
- Trinitrotoluene (TNT) and derivates
- Methyltertiarybutylether (MTBE)   … and more

Figure 3: Examples for the application of *uviblox®* systems in gas and water phase treatment for remediation and industry

Germany, elimination of odors at sludge storage in municipal waste water treatment plant (3,000 m³/h, 4.5 kW)

Germany, reduction of BTEX and CHC at disposal gas from old landfilling (250 m³/h, 6 kW)

UV-medium pressure system for degradation of TOC in highly purified water (65 m³/h, 8 kW)

Japan, elimination of odors in waste pelleting plant (6,000 m³/h, 4.5 kW)

Installation for the degradtation of TBT in waste water (60 m³/h, 2 lines à 4 × 10 kW)

4 reactors and VUV-lamps for the degradation of hydrocarbons in groundwater

IBL-*UVIBLOX*® PILOT PLANTS

A special service provided by IBL Umwelt- und Biotechnik GmbH is the set-up of pilot plants for the gas phase as well as for the liquid phase. Pilot testing is useful to find out the basic data for the dimensioning of the final treatment plant. Pilot tests usually take place on-site under real-life conditions using the medium which is to be purified. The significance of results obtained by such a pilot stage concerning scaling and upscaling is much more profound than a computer model can be.

Pilot testing is designed to support a feasibility study for the choice of technology as well as design and dimensioning of a final UV-photooxidation plant.

---

uviblox® GPT Pilot
Pilot System for Gas Phase Treatment

Technical Data

- Design: transportable compact unit including primary and secondary treatment levels
- Power rating: 1.5–12 kW (purely UV-performance)
- Flow rate: 15–250 m³/h (with integrated fan)
- Bypass connection for treatment of high flow rates (optional)
- Assembly: modular, separate reactors and container, individually activatable
- Operation: full flow, partial flow, bypass, circuit
- Dimensions (L×B×H): 2500×1200×1600 mm
- Equipment: 2× UV-reactors 1.5–6 kW
- 2× catalyst container (filling according to task)

ADVANTAGES OF *UVIBLOX®*-SYSTEMS

- Eliminating of contaminants, undesirable odours and microorganism.
- Low space requirements.
- Easily to be handled, operated, and maintained.
- Low running costs (developed form for reducing cost of energy).
- No more waste after treatment (no increase of organic carbon freight).
- Even for explosive atmospheres/gases (no open flame!).
- Suitable for a wide range of application (small and high concentration):
  $10-100,000 \, mg/m^3$
  $1-200,000 \, m^3/h$
- Adjustable reactor geometry to the given composition of water and exhaust air.

---

uviblox® WPT Pilot
Pilot System for Liqiud Phase Treatment

Technical Data

- Design: mobile compact unit on trailer
- Power rating: 2–20 kW
- Flow rate: 1–15 m³/h
  (with integrated pump)
- Bypass connection for treatment of high flow rates (optional)
- Assembly: modular, separate reactors, individually activatable
- Operation: full flow, partial flow, bypass, circuit
- Dimensions (L×B×H): 3,700×1,800×3,000 mm
- Equipment: 2 × UV-reactors 2–10 kW
- 2× dosing pump for custom-designed additives (oxidants)

- Modular design of the reactors ranging up to 20 kW power per lamp.
- Provision of a variety of modifiable UV-medium pressure lamps concerning diameter and doping which determine the spectral properties for nearly every cases.
- Different operation types concerning the internal flow of the reactor for the direct degradation of the pollutant or the degradation by adding oxidants in order to create radicals.

Figure 4: Examples for the application of *uviblox®* systems for remediation and industry

Germany, reduction of gasoline and diesel at tank cleaning

Germany, reduction of chloromethane at process gas in production of textile additives

Germany, eliminiation of odors at waste storage plant (160,000 m3/h, 75.6 kW)

CASE STUDIES

Brazil – Exhaust air and water treatment through uviblox®
Pharmaceutical Industry

The first Brazilian project was at a remediation site of a pharmaceutical company, in the city of Sao Paulo, a former production site. The soil and groundwater were polluted mainly by CHC (TRI, 1,2-DCE, VC) with concentrations max. $75,000 \mu g/l$. The remediation concept was to stop first the plume by a hydraulic barrier. So the basic design was to strip $6 m^3/h$ of contaminated groundwater and to treat the stripping gas and the stripped water by UV. The generated HCl was washed out of the gas stream by a scrubber. In the second phase the hot spots of contamination were captured by multiphase extraction. The contaminated gas stream was also included into the UV-treatment. After reducing the organic contaminants to the requested limits, the water could be used for gardening and sanitary aspects Costs were compared for the complete remediation system against traditional activated carbon. So the invest of the system with UV was I little bit higher than with activated carbon, but the operation costs are much lower. So the UV reaches the break-even-point already after 1.5 years.

A further advantage was that the UV system requires only small spaces and no big efforts for changing activated carbon onsite. The whole treatment plant was built on a part of the parking lot round a tree. The infrastructure and pre-treatment was done by GEOKLOCK. IBL's task was to design and deliver the UV components and to come to Brazil for supervision of the implementation done by GEOKLOCK.

Tropicalisation

A row of UV-systems were realised in Brazil. In some projects there appeared problems in the beginning. The solving of these problems is called »Tropicalisation«, because they are related to the special conditions in countries with tropic conditions. One aspect is that you often find groundwater with a lot of diluted iron and humic substances. Iron will cause coatings, humic substances can consume UV-light causing low efficiency. The disturbing effect of these substances can be minimised or avoided by adaption of the UV-lamps and the flow dynamics. Another point is that sometimes the energy supply at sites is not stable and/or there is a bad electrical net. Together with the fact, that even a good Brazilian net is different to a German one, some failures of components occurred. Today we adapt the electronic cabinets to these possible conditions by transforming and stabilizing the energy supply. Also the different weather conditions with high air humidity had to be taken in account. But today all these conditions can be adjusted.

Brazil – Soil air and groundwater treatment through uviblox®
Automobile industry

Problem description: In the premises of an automobile industry supplier in Sao Paulo - St. André, strong groundwater pollution with lightly volatile halogenated hydrocarbons has to be remediated. Groundwater and soil air are obtained from the subsurface by means of the multiphase extraction technique in order to be treated with two different UV facilities on-site and so achieve the pollutants removal.

Soil air treatment: The soil air treatment system comprehends a water separator and a pre-filter for the removal of particles. The main component of the treatment system is the UV reactor, where high-performance medium pressure lamps mineralize the pollutants present in the soil air ($2,000\,mg/m^3$) up to 99%. The system was design for a complete automatic operation of $500\,m^3/h$.

Groundwater treatment: For the water phase, the treatment system consists of a filtration stage to remove particles before the UV stage. This stage includes two UV reactors with medium pressure lamps which mineralize the pollutants present in the water ($55\,mg/l$) to a very low concentration values. This process takes place with the addition of an oxidant.

The system was design for a complete automatic operation of $10\,m^3/h$. The installation of stored program control elements increases the reliability, flexibility and efficiency of the reactor and the treatment process. The water is used as process water after its treatment, as well as for the sanitary installations.

Japan: Exhaust air treatment through uviblox® – Paper industry

Background: A Japanese paper company operates a plastic and paper recycling facility in the city of Fukui. The principal function of the facility is the transformation of the waste into pellets, which facilitate a stable feeding and dosage during incineration process in waste incineration plants. The compression and pellets-formation processes demand considerable temperatures that lead to the migration of high volumes of water vapour and pollutants from the waste material to the cooling air flow.

Treatment concept: This exhaust air flow must be treated before being discharge, in order to reduce the odour and pollutants emission. Therefore, the exhaust air treatment facility comprehends besides an effective aerosol, condensate and dust separation process, a photo oxidation stage for the degradation and elimination of pollutants and odours.

Implementation: The exhaust air volume to be treated is in this case $6,000\,m^3$ per hour. Concentrations of $40-50\,ppm$ Formaldehyde, $100\,ppm$ Ammonia, $50-80\,ppm$ Amine, up to $50\,ppm$ Styrene, besides other concentrations of no specified pollutants were found in the cooling air. The exigencies in the treatment are not only restricted to the pollutants concentration reduction of up to $50\,ppm$, but also include the elimination of odours.

With an installed capacity of $4.5\,kW$ are both requirements reliable and durable fulfilled. The operator currently considers the implementation of this cost-efficient treatment method in other waste incineration plants.

Germany: Landfill exhaust air treatment through uviblox®
Chemical industy

From a former landfill of a chemical company in the city of Ludwigshafen leaks exhaust soil air into the atmosphere. With direct extraction of the soil air by suction, its mean to keep the immediate buildings free of contaminated air. In the suction facility, the exhaust soil air is treated with the UV stage, which is located inside a container directly on the landfill area.

After three months of pilot test at the beginning of the project, it was concluded that this treatment method was the most cost-efficient solution in comparison to other treatment processes (e.g. activated carbon, among others). Since then is the UV treatment facility in permanent operation.

A complete and permanent measuring facilitates the registration of the pollutants concnetration in the time. The pollutants present in the soil air, cover all the pollutants spectrum contained in the landfill. The most important pollutants are BTEX (up to $400\,mg/m^3$), LHKW (up to $1,000\,mg/m^3$), methane (up to $7,000\,mg/m^3$) and organic silicon compounds (silane, siloxane). The facility is capable to treat steadily exhaust air up to $250\,m^3/h$ and reduce the concentration to less than $20\,mg/m^3$. The organic silicon compounds are completely mineralized as well and eliminated as silicon dioxide.

The extraction system comprehends a blower for the transport of the soil air and an automatic fresh-air-mixture as an explosion protection measure. The most important component of the treatment system is the UV reactor, where the pollutants from the soil air are mineralized by means of the powerful medium pressure radiator. The treatment facility has been design for a continuous operation. An active carbon filter at the end of the treatment process assures additional operational reliability. An increment of the capacity to $3,000\,m^3/h$ is planned to continue with the remediation of the affected site.

The *uviblox*® systems are applied all over the world: in Brazil, Japan, Czech Republic and numerous German companies from the sector of paint industries, pharmaceutical industries, electronic industries, land-fills (chemical industries), storage pits (odour elemination) automobile industries, paper industries and many more. Within the last few years the acceptance of this developed technology is growing step by step. After a row of successful pilot tests at remediation and industrial sites in 2005 and 2006 the demand of actual systems is growing continuously.

POTENTIAL RENEWABLE ENERGY FROM SEWAGE SLUDGE (CASE STUDY MALAYSIA)

A. Baki, M. A. Ayub, I. Atan, R. Mohd Tajuddin,
S. Ramli, J. Jaafar and S. Abdul-Talib

Faculty of Civil Engineering, Universiti Teknologi Mara,
40450 Shah Alam, Selangor, Malaysia

INTRODUCTION

As a developing nation, Malaysia is highly dependent on energy for its economic growth and contributes to the industrialization of the economy and the socio-economic improvement of the people and also the country's exports earnings. In the 9th Malaysia Plan (2006–2010), the Government will continue the efforts to promote renewable energy and energy efficiency as part of a sustainable development agenda as Malaysian progress towards Vision 2020. This initiative clearly shows that Malaysia subscribes to protecting our delicate environment while pursuing our economic development. In the Eighth Malaysian Plan (2001–2005), Renewable Energy was announced as the fifth fuel in the energy supply mix. Renewable Energy is being targeted to be a significant contributor to the country's total electricity supply. With this objective, greater efforts are being undertaken to encourage the utilization of renewable resources, such as biomass, biogas, solar and mini-hydro for energy generation (Pusat Tenaga Malaysia, 2005).

The Government has launched several economic incentives to stimulate the appearance of RE activities and technologies. Palm oil mills, sawmills, manufacturers and large institutions can start to benefit immediately by using local technology to generate income and reduce operating costs. Renewable energy resources are available in two primary forms:

(1) Waste Conversion
     a. Biomass residues from agriculture wastes
         (palm oil waste, wood waste, rice husks etc.),
     b. Municipal solid waste.

(2) Energy from the sun
     Currently several companies had already taking advantage of re-
     newable energy technologies to begin reaping energy cost savings
     and revenue in Malaysia were:
     a. TSH Bio Energy Sdn. Bhd. – As the first biomass RE project
         using empty fruit bunches as fuel, the company sold electricity
         to TNB at 21.25 sen/kwh;
     b. Jana Landfill – By producing biogas and converting it to
         electricity, this project, the first RE grid-connected project
         in Peninsula Malaysia, eventually sold power to TNB at
         16.7 sec/kwh;
     c. Bekok Kiln Drying and Moulding Sdn. Bhd. – By converting
         a fuel oil boiler to one that burns wood waste, annual fuel
         savings alone amount to RM 2 million.;
     d. Awana Kijal Golf & Beach Resort – By installing a solar water
         heating system to supply up to 35% of its consumption needs
         at a cost of RM 400,000, the resort continues to save on energy
         and maintenance and paid back its investment in only six
         years.

(Source: EIB Malaysia – 2006)

Sewage sludge

Another source that can be converted into energy is sewage sludge. Malaysia today is faced with the challenge of dramatic growth in the volume of

municipal wastewater due to rapid increase in population. Alternatives in managing the wastewater should be explored for issue to be solved. One way to achieve sustainable development is by implementing technology that converts energy from municipal wastewater sludge. Municipal wastewater contains substances used and discharged by human, such as food, beverages, pharmaceuticals, a great variety of household chemicals and substances discharged from residential areas, commercial areas and others to the sewer system. Moreover, rain water and its content also contribute to this composition. Because of its high content of organic matter, sewage sludge is used as a substrate for anaerobic digestion to recover biogas. There are about 9,500 sewage treatment plants (public and private) and over 1 million individual septic tanks in Malaysia (IWK, 2005). These sewerage facilities generated about 6.5 million tons of sewage sludge annually. All these facilities are scattered throughout the country, and thus the sludge generation is also scattered.

The management of sludge from sewage treatment plants (STPs) represents one of the major challenges in wastewater treatment today, with costs amounting to more than the treatment cost of the wastewater in many cases (Odegaard, 2004). Among several treatments, anaerobic digestion (AD) has been proven to be the most efficient technology to stabilize sewage sludge (Ray et al., 1990), especially for STPs with more than 20,000–30,000 p.e. The interest in this process has been focussed on increasing the process efficiency and reducing the investment and operational costs.

THEORETICAL REVIEW

General

Sewage sludge, also known as biosolids, is en-product after water is cleaned in waste treatment works. Gurjar (2001) describes sewage sludge as slurry with water content usually in excess of 95%. The solid phase consists principally of organic matter, derived from human, animal and food wastes.

The sewage sludge in its initial form is a liquid with 2–6% total solids (TS). On a dry basis, sludge contains 35–65% of organic matter, with the remainder being mineral noncombustible ash (Gurjar, 2001). According to Renner (2000), it is high in organic content and plant nutrients and, in theory, makes good fertilizer. It consists of 90 to 99 percent water and an accumulation of settleable solids, mainly organic that are removed during primary, secondary or advanced wastewater treatment processes but does not include grit and screenings. Sewage sludge contains significant amounts of phosphorus and nitrogen, two of the essential plant nutrients, as well as lesser quantities of heavy metals such as copper and zinc.

Sewage sludge is a potential source of phosphorus and nitrogen for use in crop production. The application of sewage sludge at a controlled rate can improve the physical and chemical properties of soils because sludge typically possesses excellent soil amendment properties (IWK, 2009).

There are different types of sludge produced from municipal wastewater treatment. Raw primary sludge is produced during the first phase of wastewater treatment. During this phase, 40–50% of the solids remove from the water (McGhee, 1991). Sludge produced from primary settling tanks is grey and has an objectionable odour. The sludge is sticky and does not drain freely. It can be successfully air-dried only in thin layers and then generally gives off foul odours (Gurjar, 2001).

Biogas

Biogas is produced by the anaerobic fermentation of organic material. Biogas production can be considered as being one of the most mature biomass technologies in terms of the numbers of installations and years of use in countries such as China and India. It has the potential for multiple uses, e.g., cooking, lighting, electricity generation, running pump sets and other agricultural machinery, and use in internal-combustion engines for motive power (Bhatia, 1990). Biogas technology is currently receiving increasing attention due to a combination of factors. Anaerobic digestion can make a significant contribution to the disposal of domestic, indus-

trial and agricultural wastes which, if untreated, could cause severe pub-
lic-health and water-pollution problems. The remaining sludge can then
be used as a fertilizer (providing there is no polluting contamination). It
therefore contributes to control of environmental hazards and recycling
of nutrients whilst alleviating dependence on imported fuels (Gunnerson
and Stuckey, 1986). When manure is used in digesters, the sludge actually performs better as a fertilizer since less nitrogen is lost during anaerobic digestion, the nitrogen is available in a more useful form, weed seeds
are destroyed, and the sludge does not smell and does not attract flies or
mosquitoes. Furthermore, it yields more useful energy than when burnt
for cooking as is the common practice in many rural regions (United Nations, 1993).

Biogas composed of methane and carbon dioxide is a by-product of anaerobic bacterial decomposition of organic waste. Municipal garbage and
sewage are important sources for biogas production (Pantskhava, 1990).
Interest in biogas as viable energy has spread throughout the globe in the
past two decades. The methane content in the biogas enables it to be used
as engine fuel and converted to heat and electricity (Baki et al., 2005).

Biogas production systems are relatively simple and can operate at small
and large scales in urban or very remote rural communities. Almost all current biogas programmes, however, are based on family-sized plants which
lose significant economies of scale, are suited more for cooking than electricity generation, and often do not produce enough output just to supply
this need. Community biogas plants are more economical and can provide
enough electricity for pumping water lighting etc. However, there are social difficulties of organization and equity in the contribution of feedstock
and the distribution of costs and benefits (United Nations, 1993).

The basic designs of biogas plants – fixed-dome (Chinese), floating-drum
(Indian), and bag (membrane) – have been used in a number of countries
for many years. The designs reflect modest optimization for reduced capital costs and increased volumetric gas yields. Biogas can be used in internal-combustion engines using either the gas alone in an adapted petrol

engine, or using a mixture of biogas and diesel in an adapted diesel engine. The main advantage of a diesel/biogas engine is the flexibility in its operation since it can operate as a dual-fuel engine using biogas and/or diesel. Usually, dual-fuel engines are so designed that when biogas is available the engine will utilize it, and when it is exhausted, the engine automatically switches over to diesel without any interruption. Diesel engines are reliable, simple to maintain, have a longer working life and higher thermal efficiency than petrol engines and are also more extensively used in rural areas. (United Nations, 1993)

Biogas technology has made some important advances in recent years, e.g. in China, Denmark and the United States. However, the technology of anaerobic digestion has not yet fully realized its promised potential for energy production. In industrialized countries biogas programmes have been hindered by operational difficulties, lack of basic understanding, and innovation. In some developing countries, development of biogas programmes has lacked urgency because of readily available and inexpensive traditional fuels such as fuel wood and residues. Lack of local skills, together with high costs, tends to be a significant deterrent to optimization and widespread acceptance of biogas technology (Hall and Rosillo-Calle, 1991).

Anaerobic Digestion

Anaerobic digestion (AD) of organic matter is a complex process carried out by a diverse microbial population and can be divided into three main stages, with a specific group of bacteria associated with each state (Hobson et al., 1981).

In the first stage, organic material, in the form of polymers, is hydrolyzed by enzymes to individual monomers, which are fermented to various intermediates including short-chain fatty acids, carbon dioxide, and hydrogen gases. The primary intermediates produced in this phase are butyric, propionic and acetic acids, with fermentation preferring the production of acetic acid (Hobson et al., 1981).

During the second stage, acetogenic bacteria metabolize butyric acid, propionic acid and other possible end products from the first stage to additional acetic acid, hydrogen and carbon dioxide. Some of the carbon dioxide and hydrogen produced by the first and second stages of this process is metabolized to acetic acid by acetogenic hydrogenatic bacteria, comprising the forth subsidiary group of bacteria (Hobson et al., 1981).

The third stage bacteria, known as methanogenic bacteria, produce biogas from the acetic acid, hydrogen and carbon dioxide by either decarboxylation of acetate or the reduction of carbon dioxide (Hobson et al., 1981). Also present in this process are sulphate reducing bacteria, which reduce sulphates and other sulphur compounds to hydrogen sulphide. Most of the hydrogen sulphide reacts with heavy metal salts, including iron, to form insoluble sulphides, but there will always be some hydrogen sulphide in the biogas (Rivard and Boone, 1995). In dairy biogas, the amount of hydrogen sulphide in dairy biogas usually ranges from 1,000 to 5,000 ppm.

As Figure 1 demonstrates, there are two main routes by which methane is produced during the process of anaerobic digestion. Methane may be derived by the decarboxylation of acetate to methane and carbon dioxide or by the reduction carbon dioxide to methane. Most methanogens, or methane-producing bacteria, have the metabolic capacity to grow in the absence of oxygen by oxidizing hydrogen and reducing carbon dioxide to methane. In this case, hydrogen is the energy source (electron donor) and carbon dioxide is the electron acceptor (Hobson and Richardson, 1985). Research has shown that 73% of the methane produced is derived from the methyl group of acetate (Jones et al., 1980). It is likely that the fermentation of carbohydrates is biased towards the production of acetic acid. Thus:

$$2\ C_6H_{12}O_6 + H_2O \ \rightarrow \ CH_3COOH + 2\ CO_2 + 4\ H_2$$

With the formation of acetic acid, carbon dioxide and hydrogen, methane is then formed by the two following pathways:

$$2\ CH_3COOH \rightarrow 2\ CH_4 + 2\ CO_2$$
$$4\ H_2 + CO_2 \rightarrow CH_4 + 2\ H_2O$$

From these processes, approximately 66% of the methane comes from acetic acid and 33% from the hydrogen. There are three main factors which contribute to the predominance of acetic acid as a precursor of methane. Because the hydrogen formed is used to hydrogenate the long-chain fatty acids, to reduce sulphur and nitrogen compounds and to form bacterial cellular components, hydrogen may not be available for methane formations. Second, because animal wastes always contain volatile fatty acid (VFA) from fermentation that occurs in the animal's stomach, VFA concentrations may be high. Acetic acid is a large constituent of this acid mixture (Hobson and Richardson, 1985).

Finally, when the second group of bacteria (known as the acetogenic hydrogenating bacteria) convert hydrogen and carbon dioxide to acetate, the proportion of methane generated from acetate increases (Hobson and Richardson, 1985).

Figure 1: Anaerobic Digestion Process (From Hobson and Richardson, 1985)

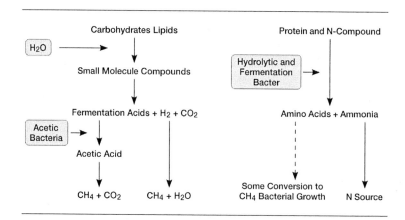

Digestion Process

Biogas microbes consist of a large group of complex and differently acting microbe species, notable the methane-producing bacteria. The whole biogas-process can be divided into three steps: hydrolysis, acidification, and methane formation (Figure 2).

This is a complex physio-chemical and biological process involving different factors and stages of change. This process of digestion (methanization) is summarized below in its simple form. The breaking down of inputs that are complex organic materials is achieved through three stages as described below:

Stage 1: Hydrolysis

The waste materials of plant and animal origins consist mainly of carbohydrates, lipids, proteins and inorganic materials. Large molecular complex substances are solubilized into simpler ones with the help of extracellu-

**Figure 2: Three-Stage Anaerobic Fermentation of Biomass
(from Rivard and Boone, 1995)**

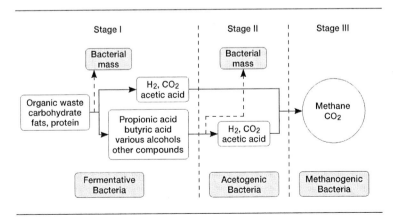

lar enzyme released by the bacteria. This stage is also known as polymer breakdown stage. For example, the cellulose consisting of polymerized glucose is broken down to dimeric, and then to monomeric sugar molecules (glucose) by cellulolytic bacteria. During this phase the digestion of digestion the decomposable solids, required mainly for bacteria food, include sugar, starch, cellulose and soluble compound of nitrogen (i. e. nitrites and nitrates). Anaerobic bacteria use oxygen from organic matters and the soluble compounds of nitrogen (Gurjar, 2001).

Stage 2: Acidification

The monomer such as glucose which is produced in Stage 1 is fermented under anaerobic condition into various acids with the help of enzymes produced by the acid forming bacteria. At this stage, the acid-forming bacteria break down molecules of six atoms of carbon (glucose) into molecules of less atoms of carbon (acids) which are in a more reduced state than glucose. The principal acids produced in this process are acetic acid, propionic acid, butyric acid and ethanol. Soluble organic component the products of hydrolysis are converted into organic acids, alcohols, hydrogen and carbon dioxide by acidogens (Li and Noike, 1992).

Stage 3: Methanization

The principle acids produced in Stage 2 are processed by methanogenic bacteria to produce methane. The reaction that takes place in the process of methane production is called Methanization and is expressed by the following equations (Karki and Dixit, 1984).

$CH_3COOH \rightarrow CH_4 + CO_2$
Acetic acid   Methane   Carbon dioxide

$2 CH_3CH_2OH + CO_2 \rightarrow CH_4 + 2 CH_3COOH$
Ethanol   Carbon dioxide   Methane   Acetic acid

$CO_2 + 4 H_2 \rightarrow CH_4 + 2 H_2O$
Carbon dioxide   Hydrogen   Methane   Water

The above equations show that many products, by-products and intermediate products are produced in the process of digestion of inputs in an anaerobic condition before the final product (methane) is produced. Obvi-
ously, there are many facilitating and inhibiting factors that play their role
in the process. Some of these factors are discussed below:

pH value

The optimum biogas production is achieved when the pH value of input
mixture in the digester is between 6 and 7. The pH in a biogas digester is
also a function of the retention time. In the initial period of fermentation,
as large amounts of organic acids are produced by acid forming bacteria,
the pH inside the digester can decrease to below 5. This inhibits or even
stops the digestion or fermentation process. Methanogenic bacteria are
very sensitive to pH and do not thrive below a value of 6.5. Later, as the
digestion process continues, concentration of NH_4 increases due to digestion of nitrogen which can increase the pH value to above 8. When the
methane production level is stabilized, the pH range remains buffered between 7.2 to 8.2. (Karki and Dixit, 1984).

Temperature

The methanogens are inactive in extreme high and low temperatures.
The optimum temperature is 35 degrees C. When the ambient temperature goes down to 10 degrees C, gas production virtually stops. Satisfac-
tory gas production takes place in the mesophilic range, between 25 degrees to 30 degrees C. Proper insulation of digester helps to increase gas
production in the cold season. When the ambient temperature is 30 degrees C or less, the average temperature within the dome remains about
4 degrees C above the ambient temperature (Lund et al., 1996).

Retention Time

Retention time (also known as detention time) is the average period that
a given quantity of input remains in the digester to be acted upon by the

methanogens. In a cow dung plant, the retention time is calculated by dividing the total volume of the digester by the volume of inputs added daily. Considering the climatic conditions of Nepal, a retention time of 50 to 60 days seems desirable. Thus, a digester should have a volume of 50 to 60 times the slurry added daily. But for a night soil biogas digester, a longer retention time (70–80 days) is needed so that the pathogens present in human faeces are destroyed. The retention time is also dependent on the temperature and up to 35 degrees C, the higher the temperature, the lower the retention time (Lagrange, 1979).

Anaerobic Treatment Systems for Municipal Wastewater

If anaerobic processes could be shown to treat dilute wastewater consistently and reliably, it would be a highly significant development in wastewater treatment. Since anaerobic fermentation results in a lower cellular yield, less sludge is generated, and hence lower sludge handling costs would be possible. In addition, lower energy requirements would result, since aeration would not be necessary, and methane would be produced as a by-product. In fact, the treatment of wastewater might be a net energy producer (Switzenbaum and Jewell, 1978).

Originally, anaerobic treatment was the preferred process for domestic wastewater management. Imhoff modified the septic tank for wastewater treatment in Germany, and by 1933 the Imhoff tank was used by over 240 towns in Germany. In general, these early processes were poor for removal of soluble BOD but were successful in capturing solids. Thus the anaerobic processes were abandoned, in practice, for liquid municipal wastewater treatment, with the development of stricter effluent standards and, until the middle part of the 1970s, the anaerobic fermentation process was not considered practical for treating low strength wastewater (BOD < 500 1,000 mg/l) (Marchaim, 1992).

Biogas Conversion

Biomass or biogas can be converted into electricity in one of several processes. The majority of biomass electricity is generated today using a steam cycle, as shown in Figure 2. In this process, biomass is burned in a boiler to make steam. The steam then turns a turbine, which is connected to a generator that produces electricity (EERE, 2006). From a thermodynamic point of view the steam turbine occupies a favourable position, as it can translate into mechanical work a relatively large fraction of the heat energy rendered available the expansion of the steam in the turbine. Its thermal economy is also fairly good, especially in the turbines of large output and operating at fairly high pressures (Kearton, 1958).

Converting biogas into electricity can also be established by utilizing the pressure of biogas itself through turbo generator. Consequently, the potential of biogas pressure has yet to be further formative, since that electricity can be solely produced by biogas pressure. In the future, fuel cells and the Stirling engine may be able to use biogas to cost-effectively generate electricity and recover process heat (IEA Bioenergy, 2001).

Pressure of Biogas

Mixing is regarded to be essential in high-rate digesters. Mechanical agitation, recirculation of digester contents or recirculations of produced gas are currently used as the practical mixing methods (David et al., 1980; Sang et al., 1995). Among these mixing methods, gas recirculation appears to be the most efficient (Konstandt and Roediger, 1977; Monnet, 2003; Sang et al., 1995). However, the gas pumping system for compressing and releasing the gas into the digester is expensive to operate. To reduce the cost of pumping, use of high pressure, created by the production of gas during anaerobic digestion, as a power source can be suggested to replace a mechanical power source.

Sang et al. (1995) reported that a pilot-scale digester employing a unique mixing method of gas recirculation, which uses the pressure of produced gas as the source of mixing power, was constructed and operated successfully. The relationship between fermentor pressure and gas production rate is shown in Figure 3.

Figure 3: Effect of Fermentor Pressure on the Gas Production Rate
During Anaerobic Digestion. (Sang et al., 1995).

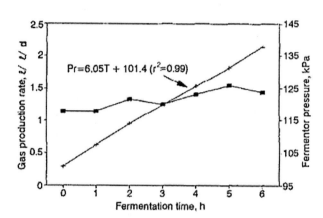

Codes in Figure: Pr: Fermentor Pressure [kPa]; T: Fermentation Time [h].

The fermentor pressure increased linearly with increasing fermentation
time (Pressure [kPa] = 101.4 + 6.05 × Time [h], r^2 = 0.99). The highest
pressure of 140 kPa achieved in this experiment is the same as the hydraulic pressure of the bottom zone of a digester 4 m in height. It would therefore be possible to use the gas pressure achieved during the fermentation
as mixing power (Sang et al., 1995).

Changes of pressure and temperatures in the digester are shown in Figure 4. While the temperatures showed no marked change, remaining at
around 39°C, the pressure showed a cyclic change with 6 h intervals. The
pressure in the digester reached atmospheric pressure immediately after
gas diffusion and 114.5 kPa within 70 min, remaining at this level thereaf-
ter until the next diffusion.

One approach was to increase the digester pressure to 206.7 kPa has been
reported (Hayes et al., 1990). Since carbon dioxide is 40 times more soluble in water than methane, such a high digester pressure results in super

Figure 4: Changes of Temperatures and Pressure in Pilot-Scale Digester in One Day (Sang et al., 1995).

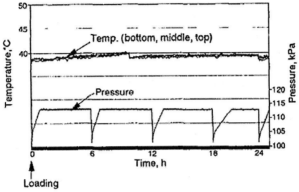

Codes in Figure: Pr: Fermentor Pressure [kPa]; T: Fermentation Time [h].

saturation of carbon dioxide in the digester and therefore production of gas with a considerably high methane content (Sang et al., 1995).

CASE OF MALAYSIA

The management of sludge in Malaysia differs from many developed countries. Sludge generated and managed at about 10,000 sewage treatment plants around the country (IWK, 2009). The sizes of the sewage treatment plants are generally small, mostly less than 20,000 PE. Thus, conventional renewable energy initiatives are not financially feasible due to logistics problems of diversely located small size sewage treatment plants (Baki et al., 2006a). Therefore, alternative methods of renewable energy generation for small scale sewage treatment plants are needed in the case of Malaysia (Baki et al., 2006b).

In order to determine basic relationship between sludge and biogas at

small scale, experiments need to be conducted. Experiments were conducted at bench scale in laboratories to determine relationships between
sludge volume and biogas volume; between organic content and methane
generation; and between organic content and pressure of biogas. Sludge
samples were collected from sewage treatment plants within and within
vicinity of Universiti Teknologi MARA, Shah Alam campus, Malaysia.
Chemical parameters like BOD_5, COD and SS were tested using the
Standards Method (APHA et al., 2006).

The analysis on organic contents represented by biological oxygen demand (BOD_5) and suspended solid (SS) was conducted before and after
the experiments to indicate the removal organic of content during 40 days
of experiment. However, the chemical oxygen demand (COD) values increases instead of the expected decreases. This may be due to changes in
temperature and pH of the sludge.

Methane generation was observed during the experiment. However, the
relationship between organic content and methane generation cannot be
clearly defined because of the inconsistency observed between the three
different parameters, BOD_5, COD and SS. Based on literatures, higher
organic content will result in higher the methane generation. It means
that higher removal on BOD_5, COD and SS will contribute to higher
methane generation. But in this study, the COD has increased instead of
the expected decrease due to organic removals.

The organic contents of sewage sludge were increased as increase of fermentation time. High in BOD_5, COD and SS concentrations indicate
possibility of high in organic carbon thus resulting in inhibitory of meth-
ane-forming bacteria (methanogens). For this reason, instead of being increase, the pressure of biogas decreased and even illustrated the negative
values. The inhibition of the biogas process occurred with increasing carbon dioxide production and its utilization which particularly correspond-
ence to high organic carbon and high concentrations of BOD_5 and COD
and in the end resulting in negative pressure of biogas. This scenario has
been discussed by Gerardi (2006) where increase in carbon dioxide pro-

duction can occur through increased organic loading to the digester or decrease methanogenesis.

From the experiments conducted, it was found that different volume of sludge will produced different volume of biogas depends on the characteristics of the sludge. The relationship between sludge volume and the volume of biogas can be determined. If the volume of sludge volume increased, the production of biogas will also increase. Thus, both sludge volume and biogas have proportional relationship. However this proportional relationship only applies up to a number of days when the volume of biogas will remain constant.

CONCLUDING REMARKS

This study has achieved its objectives, which include determining the relationship between sludge volume and biogas volume; between organic content and methane generation; and between organic content and pressure of biogas. These bench scale relationships will enable scaled models to be developed for further studies on the relationship between sludge and biogas, which in turn will enable assessment of potential renewable energy from the sewage sludge.

REFERENCES

APHA, AWWA and WEF (2006),
Standard Methods for the Examination of Water and Wastewater,
The American Public Health Association,
The American Water Works Association
and The Water Environment Federation, 21st Edition.

Baki, A., Abdul-Talib, S., Abdul Hamid, M. H., and Khor B. C., (2005),
Energy from sewage sludge: Potential Application in Malaysia,
Proceedings of 3rd Workshop on Regional Network Formation for

Enhancing Research and Education on Green Energy Technologies,
p63, Batu Ferringhi, Penang, 7–10 August 2005.

Baki, A., Abdul-Talib, S., Abdul Hamid, M. H.,
Khor, B. C., Salleh, M. T. (2006a),
The Prospect of Energy from Sewage Sludge in Malaysia,
Proceedings of the International Conference on Energy
for Sustainable Development: Issues and Prospect for Asia,
1–3 March 2006, Phuket, Thailand.

Baki, A., Abdul-Talib, S., Abdul Hamid, M. H.,
Khor, B. C., Abdul Jabbar, A. R. (2006b),
*Renewable Energy using Biogas from Sewage Sludge: A Case Study of
Potential at Shah Alam Sewage Treatment Plant, Malaysia*,
ORBIT 2006: 5th International Conference on Biological Waste
Management, 13th to 15th Sept. 2006, Weimar, Germany.

Bhatia, R. (1990),
*Diffusion of renewable energy technologies in developing countries:
a case study of biogas engines in India*, World Development,
Vol. 18 (2006) 575–590, Energy Information Bureau (EIB) Malaysia:
http://eib.ptm.org.my (access on 07/09/2008)

David, A. S., Dannis, H., and Rex, H., (1980),
Designing a digester, 73–112, in: Stafford, D. A. (ed.), Methane
production from waste organic matter, CRC Press, Boca Raton.

EERE (2006),
Biomass Power,
http://www.eere.energy.gov/de/biomass_power.html
(access on 06/11/2008)

Energy Information Bureau (EIB) Malaysia (2006),
Energy Policy & Planning.
http://eib.ptm.org.my (accessed on 20/07/2009)

Gerardi, M. H., (2006),
Wastewater Bacteria, John Wiley and Sons, New Jersey.

Gunnerson, C. G and Stuckey, D. C. (1986),
Integrated resource recovery: aerobic digestion – principles and practices for biogas systems, World Bank Technical Paper, 49, 155ff,
World Bank, Washington USA.

Gurjar, B. R., (2001),
Sludge treatment and disposal, A. A. Balkema, Tokyo.

Hall, D. O., and Rosilto-Calle, F. (1991),
Biomass energy resources and policy (biomass in developing countries),
report to the United States Congress Office of Technology Assessment
(Energy in Developing Countries) (unpublished).

Hayes, T. D., Issacson, H. R., Pfeffer, J. T., and Liu, Y. M., (1990),
In situ methane enrichment in anaerobic digestion,
Biotechnology, Bioengineering, 35, 73 – 86.

Hobson, P. N., Bousfield, S. and Summers, R. (1981),
Methane Production from Agricultural and Domestic Waste.

Hobson, P. N. and Richardson, A. J. (1985),
The Microbiology of Anaerobic Digestion, National Institute for
Research, in: Dairyin, Technical Bulletin, Volume 1528.

IEA Bioenergy (2001),
Biogas and more! – Systems and Markets Overview of Anaerobic digestion,
AEA Technology Environment, Culham, Abingdon, Oxfordshire, UK.
http://www.iea.com/biogas (access on 26/08/2008)

Indah Water Konsortium Sdn. Bhd. (2005),
Revised Sludge Strategy, unpublished.

Indah Water Konsortium Sdn. Bhd. (2009),
http://www.iwk.com.my (IWK website, accessed on 20/072009)

Jones, D. D., John, C. N., Alvin, C. D. (1980),
Methane Generation from Livestock Waste, Publication #AE-105,
Purdue University Cooperative Extension.

Karki, A. B. and Dixit K. (1984),
Biogas Fieldbook, Sahayogi Press, Kathmandu, Nepal.

Kearton, W. J., (1958),
*Steam Turbine Theory and Practice – A Textbook for Engineering
Students*, Pitman Publishing, New York, 7th Edition.

Konstandt, H. G. and Roediger, A. G., (1977),
Engineering operation and economics of methane gas production,
in: Schlegel, H. G. and Barnea, J. (ed.), Microbial energy conversion,
Pergamon Press, New York, 379–392.

Lagrange, B. (1979),
Biomethane 2: Principles – Techniques Utilization,
EDISUD, La Calade, 13100 Aix-en-Provence, France.

Li, Y., and Noike, T., (1992),
*Upgrading of anaerobic digestion of waste activated sludge by thermal
pretreatment,* Water Science Technology, 26, 857–866.

Lund, M. S., Andersen, S. S. and Torry-Smith, M. (1996),
Building of a Flexibility Bag Biogas Digester in Tanzania,
Student Report, Technical University of Denmark, Copenhagen.

Marchaim, U. (1992),
Biogas processes for sustainable development,
MIGAL Galilee Technological Centre Kiryat Shmona.

McGhee, T.J. (1991),
Water Supply and Sewerage, 6th Edition, McGrawHill.

Monnet, F., (2003),
An introduction to anaerobic digestion of organic wastes,
http://www.yahoo.com/anaerobicdigestion/Monnet
(accessed on 26/02/2008)

Odegaard, H., 2004.
Sludge minimization technologies — an overview,
Water Science Technology 49 (10), 31–40.

Pantskhava, E.S., (1990),
Biomass and the Problems of Ecology, Agrochemistry and Energy,
Environmental Biotechnology, Elsevier Science Publishing Company
Inc, NY.

Pusat Tenaga Malaysia (2005),
Option and Tools for Renewable Energy Implementation in Malaysia,
2nd Seminar on Harmonised Policy Instruments for the Promotion
of Renewable Energy and Energy Efficiency in the ASEAN Member
Countries, Glenmarie, 06–07/09/2005

Ray, B.T., Lin, T.G., Rajan, R.V., (1990),
Low level alkaline solubilization for enhanced anaerobic digestion,
Res. J.Water Pollut, Control Fed 62, 81–87.

Renner, R., (2000),
Sewage sludge, pros & cons,
Environmental Science & Technology, 34, 19.

Rivard, C. and Boone, D. (1995),
The Anaerobic Digestion Process, Second Biomass Conference of
the Americas, 1995 – Portland Oregon: National Renewable Energy
Laboratory.

Sang, R. L., Nam, K. C., and Won, J. M., (1995),
*Using the pressure of biogas created during anaerobic digestion as the
source of mixing power*, Journal of Fermentation and Bioengineering,
8, 4, 415–417.
http://www.sciencedirect.com/pressurebiogas (accessed on 04/03/2008)

Switzenbaum, M. S. and Jewell, W. J., (1978),
Anaerobic attached film expanded bed reactor treatment of dilute organics,
in: Proceedings of the 51st Annual WPCF Conference, 1–164.

United Nations (1993),
Nairobi (Habitat): Application of Biomass Energy Technologies,
Centre for Human Settlements, p168.

HEAVY METAL HYPERACCUMULATING PLANTS IN MALAYSIA AND ITS POTENTIAL APPLICATIONS

Marcus Jopony and Felix Tongkul

School of Science and Technology, Universiti Malaysia Sabah
Locked Bag No. 2073, 88999 Kota Kinabalu, Sabah, Malaysia

ABSTRACT

Metal hyperaccumulating plants are plants that possess unique ability to accumulate up to 10,000 ug/g or higher (dry matter) of specific heavy metals from the soil without any apparent damage to the plant. Out of approximately 420 metal hyperaccumulators that have so far been identified worldwide, less than 20 species were from Southeast Asia. Five of these species including *Rinorea bengalensis (Violaceae)* and *Dichapetalum gelonioides (Dichapetalaceae)* can be found in Malaysia. *R bengalensis* is a strong Ni hyperaccumulator and analysis of field specimens from Sabah, Malaysia showed the average Ni concentration in its leaves is about 1.2% (12,000 ug/g) dry matter, which is about 5× compared with the natural Ni concentration (2,200 ug/g) in the soil. The metal is also hyperaccumulated in other above ground tissues of the plant. *D. gelonioides* is also strong hyperaccumulator of Ni as well as Zn. However, further studies on this species are presently hampered by difficulty in obtaining field specimens. The above plants partially fulfill the criteria required for a candidate plant in phytoremediation and phytomining technologies. The existence of such indigenous metal hyperaccumulating plants therefore provides opportunities for R & D, including in the areas of plant genetics and biotechnology, on the above technologies in Malaysia as well as in this region.

INTRODUCTION

Heavy metal hyperaccumulating plants (i.e. metal hyperaccumulators) are plants that possess unique ability to accumulate 1,000 – 10,000 mg/kg (dry matter) or higher of specific heavy metals in their leaves without any apparent damage to the plant (Brook et al., 1977a). Some 420 species of these unique plants are known worldwide and approximately 75% are nickel hyperaccumulators (Reeves & Baker, 2000). Most of these plants are rare and endemic to metalliferous soils.

Scientific studies or reports on these unique plants are mainly concentrated on few genera of European or temperate species, namely *Alyssum (Brassicaceae)*, *Thlaspi (Brassicaceae)* and *Berkheya (Asteraceae)*. These species can accumulate concentration of Ni in excess of 2% on a dry matter basis (Brooks and Radford, 1978; Reeves & Brooks, 1983b). By contrast, information on metal hyperaccumulators from Southeast Asia is still limited (e.g. Visoottiviseth et al., 2002; Reeves, 2002; Jopony & Baker, 2000; Baker et al., 1992; Brooks, 1987). These findings were, however, mostly based on analysis of herbarium materials (i.e sterile specimens) rather than on field collections. Despite this region being noted for it's diverse flora, the number of metal hyperaccumulators discovered and identified so far is less than 20, and five of them can be found in Malaysia (Table 1). The most prominent metal accumulators from Southeast Asia, including in Malaysia, are *Rinorea bengalensis* and *Dichapetalatum gelonioides* (Jopony & Baker, 2000). This paper describes these two unique plants, including its potentials in phytoremediation and phytomining technologies.

MATERIALS AND METHODS

Samples of above ground tissues (trunk bark, twig bark, leaves and fruits) of randomly selected *R. bengalensis* plants were collected from two ultrabasic areas in Sabah, Malaysia. Surface (0 – 20 cm) soil samples were also collected adjacent to each plant specimen. The plant samples were thoroughly washed in running water, rinsed in distilled water and oven-dried (105°C). Soil samples were air-dried and sieved (< 2 mm). For determination of total metal content, 0.5 – 1.0 g of plant tissue or soil sample was digested in 20 ml hot concentrated nitric acid (AR grade) until the final volume of the digest solution was less that 5 ml. For comparison, a separate extraction in hot distilled water was also carried out. After cooling, the acid digests and hot water extracts were partially diluted, filtered (Whatman #541) and the filtrate made up to 100 ml volume. Analysis of Ni and other selected elements in the filtrate was done by atomic absorption spectrophotometry with reference to standard AAS procedures. All concentration data were expressed on a dry-weight basis. Due to difficulty in getting field specimens, information on *D. gelonioides* are derived from herbarium records and available secondary data.

RESULTS AND DISCUSSIONS

R. bengalensis and D. gelonioides in Sabah, Malaysia

R. bengalensis is a widespread species in Southeast Asia, and in Malaysia it is supposed to be found in primary lowland and hill forests (Ng, 1995). This plant, however, is becoming a relatively rare species due to conversions of primary forests into other types of landuse. Except in very remote areas, most of the localities listed in herbarium collections are no longer primary forests. The present study carried out in Sabah, Malaysia, managed to find some localities with ultrabasic subtrate where field specimens of *R. bengalensis* can be found.

D. gelonioides is also reported to be widely distributed in Southeast Asia, including in Sabah, Malaysia. However, contrary to herbarium records, attempts to locate field specimens of *D. gelonioides*, were so far unsuccessful. Again, this can be attributed to losses of natural habitats. Some characteristic features of both plants are shown and compared in Table 2.

Table 1:   List of metal hyperaccumulators from Southeast Asia

Ref.	Location/ Origin	FAMILY/ Species	Metal/ Foliar metal*
Brooks & Wither. 1977	Southeast Asia	VIOLACEAE Rinorea bengalensis	Nickel (Ni); 17,500 mg/kg
Brooks et al. 1977b	Southeast Asia	VIOLACEAE Rinorea javanicas	Nickel (Ni); 2,170 mg/kg
Wither & Brooks 1977	Southeast Asia	SAPOTACEAE Planchonella oxyedra	Nickel (Ni); 19,600 mg/kg
Wither & Brooks 1977	Indonesia	TILLIACEA Trichospermum kjellbergii	Nickel (Ni); 3770 mg/kg
Baker et al. 1992	Sumatra, Indonesia	DICHAPETALACEAE Dichapetalatum gelonioides ssp. tuberculatum	Zinc (Zn) 30,000 mg/kg
Baker et al. 1992	Kalimantan, Indonesia	DICHAPETALACEAE Dichapetalatum gelonioides ssp. sumatranum	Zinc (Zn) 14,000 mg/kg
Reeves, R. D. 2002	Indonesia;	EUPHORBIACEAE Phyllantus insulae-japen	Nickel (Ni); 38,720 mg/kg
Reeves, R. D. 2002	Sulawesi, Indonesia;	EUPHORBIACEAE Glochidion aff. acustylum	Nickel (Ni); 6,060 mg/kg
Reeves, R. D. 2002	Sulawesi, Indonesia;	VIOLACEAE Rinorea bengalensis	Nickel (Ni); 17,350 mg/kg
Baker et al. 1992	Mindanao, Philippines	EUPHORBIACEAE Phyllanthus securinegoides	Nickel (Ni); 34,750 mg/kg
Baker et al. 1992	Palawan, Philippines	EUPHORBIACEAE Phyllanthus »palawanensis«	Nickel (Ni); 16,230 mg/kg
Baker et al. 1992	Mindanao, Philippines	DICHAPETALACEAE Dichapetalatum gelonioides ssp. pilosum	Zinc (Zn) 26,360 mg/kg
Baker et al. 1992	Philippines	DICHAPETALACEAE Dichapetalatum gelonioides ssp. Tuberculatum	Nickel (Ni) 25,820 mg/kg
Baker et al. 1992	Palawan, Philippines	MELIACEAE Walsura manophylla	Nickel (Ni) 7,090 mg/kg
Baker et al. 1992	Palawan, Philippines	OCHNACEAE Brackenridgea palustris ssp foxworthyi	Nickel (Ni) 7,600 mg/kg
Vissottiviseth et al. 2002	Thailand	PARKERIACEAE Pityrogramma calomelanos	Arsenic (As) 8,000 mg/kg
Vissottiviseth et al. 2002	Thailand	PTERIDACEAE Pteris vittata	Arsenic (As) 6,030 mg/kg
Proctor et al. 1989	Sabah, Malaysia	DIPTEROCARPACEAE Shorea tenuiramulosa	Nickel (Ni) 1,000 mg/kg
Baker et al. 1992	Sabah, Malaysia	EUPHORBIACEAE Phyllanthus lamprophyllus	Nickel (Ni); 9,210 mg/kg

Ref.	Location/ Origin	FAMILY/ Species	Metal/ Foliar metal*
Baker et al. 1992	Sabah, Malaysia	DICHAPETALACEAE Dichapetalatum gelonioides ssp. sumatranum	Zinc (Zn) 15,660 mg/kg
Baker et al. 1992	Sabah, Malaysia	DICHAPETALACEAE Dichapetalatum gelonioides ssp. Tuberculatum	Nickel (Ni) 26,650 mg/kg
Reeves, R. D. 2002	Sabah, Malaysia	RUBIACEAE Psychotria cf. gracili	Nickel (Ni) 10,590 mg/kg

* denotes highest concentration recorded

Table 2:   Some characteristic features of *R. bengalensis* and *D. gelonioides* plant

	Description	
Feature	R. bengalensis	D. gelonioides
Family	Violaceae	Dichapetalaceae
Genus	Rinorea	Dichapetalum
Species	Rinorea bengalensis	Dichapetalum gelonioides
Sub-species		(i) spp. tuberculatum (ii) ssp sumatranum (iii) ssp. pilosum
Distribution	Southeast Asia	Southeast Asia
Locality	Sabah, MALAYSIA	Sabah, MALAYSIA
Soil type	Ultrabasic (Serpentine)	Ultraabasic & Non-ultrabasic
Forest type	Primary forest	Primary forest
Plant type	Shrub/Small tree up to 5 m tall	Ultraabasic & non-ultrabasic
Leaves (size)	Medium to large up to 45 cm long and 18 cm wide)	Small to medium size
Leaves (dry matter)	Up to 5.0 g per leaf (Average ~ 2.5 g/leaf)	Low biomass
Fruits	Round (12 × 12 mm) Green pea like; Up to 3 seeds/fruit	Small yellowish (ssp tuberculatum)

Table 3:   Average metal concentration in leaves and trunk barks of field specimens
            (n = 6) of R. bengalensis from ultrabasic areas in Sabah, Malaysia

	Metal Concentration (mg/kg DW)						
	Ni	Zn	Co	Cr	Cu	Fe	Mn
Leaves	12,010	66	< 10	< 10	< 10	162	171
Bark (Trunk)	12,717	390	< 10	< 10	< 10	281	68
Soil	2,234	63	356	440	36	12.80*	4,424
Leaves[1]	24	45	–	–	–	–	–
Leaves[2]	34	20	–	–	–	–	–

* denotes concentration expressed as percent dry weight;
[1] Litsea sp. (laureceae): non-hyperaccumulator;
[2] Eugenia sp. (Myrtaceae): non-hyperaccumulator.

Foliar Metal Content of R. bengelensis and D. geloniodes

The average foliar Ni content obtained for six randomly selected field specimens of R. bengalensis plant is about 12,000 mg/kg (or 1.2%). This figure is about 5× higher than the natural Ni concentration in the soil, and
about 400× higher compared with the foliar Ni of two common ultrabasic flora, namely Litsea sp. and Eugenia sp., collected from the study area
(Table 3). This result support the fact that R. bengalensis is a hyperaccumulator of Ni when growing over ultrabasic rocks (Brooks & Wither, 1977).

The foliar Ni for the field specimens also falls within the range
836–17,500 mg/kg reported by Brook & Wither (1977) for 21 herbarium specimens, and of the same order of magnitude as with many other
known nickel hyperaccumulators (Reeves & Brooks, 1983; Brooks & Radford, 1978; Jaffre et al., 1976). R. bengalensis seems to accumulate zinc
to a certain degree (especially in the trunk bark) while other metals (i.e
Co, Cr, Cu, Fe and Mn) are very low. The Co/Ni ratio is < 0.01, supporting the fact that the plants were derived from an ultrabasic substrate
(Brooks & Wither, 1977).

An average of 40% of the total Ni in the above-ground tissues of
R. *bengalensis* can be extracted using hot distilled water, indicating that
the Ni in the tissues is partially present as fairly soluble compounds.
By contrast, less than 1.0% of the total soil Ni is water-extractable. Despite the low solubility of soil Ni, the plant was capable of absorbing Ni
through its roots and subsequently hyperaccumulate it in above ground
tissues.

Depending on its sub-species, D. *gelonioides* is a strong Ni or Zn hyperaccumulator. The highest foliar Ni for sterile specimens of *ssp. tubercu-
latum* from Sabah, Malaysia is 26,650 mg/kg while foliar Zn for *ssp. su-
matranum* is 15,660 mg/kg (Baker et al., 1992). These figures are higher
than those obtained for R. *bengalensis* (Table3).

Distribution of Ni, Ca and Mg in R. bengalensis

The Ni content of various organs of R. *bengalensis* is shown in Table 4.
The data was based on two field specimens. Nickel is evenly hyperaccumulated from the trunk to the leaves tissues while relatively lower
concentration (but still above 1,000 mg/kg) can be found in the fruit.
Higher Ni accumulation seems to occur in the older tissues such as
lower trunk and older leaves. The distribution of Ni in the Malaysian
specimens of R. *bengalensis* is comparable with several other hyperaccumulators (Table 4).

It is commonly regarded that the infertility of ultrabasic soils is associated with the unfavourably low Ca/Mg mole quotient and high Mg
(Brooks, 1987). As expected, for the soils studied the concentration of Mg
> Ca and the Ca/Mg quotient < 1 (Table 5). Nevertheless, R. *bengalen-
sis* seemed to be unaffected by the above factors. In fact this plant species can efficiently accumulate remarkably high Ca concentration in the
above ground tissues (Table 5). As with Ni, Ca concentrations decrease
from the trunk to the fruit.

Table 4:   Nickel content (%) of various organs of *R. bengalensis* (n = 2) compared
          with values reported for three other hyperaccumulators

	Present study		Severne & Brooks 1972	Jaffre & Schmid 1974	Jaffre et. al. 1976
Species	*Rinorea bengalensis*		*Hybanthus floribundus*	*Psychotria douarrei*	*Sebertia accuminata*
Locality	Sabah, Malaysia		Western Australia	New Caledonia	New Caledonia
Organ	S1	S2			
Fruit (skin)	0.46	0.32	0.31	2.30	0.30
Fruit (seed)	0.58	0.53			
Leaves (young)	1.34	1.12	0.71	3.40	1.17
Leaves (old)	1.54	1.12			
Twigs (bark)	1.15	1.13		5.52	1.12
Upper trunk (bark)	1.34	1.15	0.17	5.24	2.45
Lower trunk (bark)	1.50	1.16			

Potentials of R. bengalensis and D. gelonioides for Phytoremediation and Phytomining

In the past, metal hyperaccumulators are of interest partly of their possible significance in mineral exploration or prediction of rock type, and partly because of the interesting problems in plant chemistry and plant physiology associated with high accumulations of metal, which is normally phytotoxic to vegetation (Brooks & Wither, 1977). However, current interests in metal hyperaccumulators focus on its potential utilization in cleaning / decontamination of metal-polluted soils @ phytoremediation (Baker et al., 2000; Raskin et al., 1997; Robinson et al., 1997a; Salt et al., 1995; Baker et al., 1994) and for extraction of metals from low-grade ore body or mineralised soil @ phytomining (Anderson et al., 1999; Brooks et al., 1998; Nick & Chambers, 1998; Chaney et al., 1998; Robinson et al., 1997a).

Table 5:   Ca and Mg content, and Ca/Mg quotient of *R. bengalensis*

Organ	Ca (%)	Mg(%)	Ca/Mg **
Fruit (skin)	1.41	0.11	7.86
Fruit (seeds)	2.51	0.33	4.58
Leaves (young)	2.61	0.43	3.64
Leaves (old)	2.94	0.39	4.53
Twigs (bark)	7.21	3.82	27.22
Upper Trunk (bark)	7.19	1.58	32.87
Lower Trunk (bark)	9.20	1.69	27.29
Soil	1.12	2.15	0.31

** Quotient expressed in terms of moles

The candidate plant for the above environmentally friendly and relatively inexpensive technologies need to fulfil certain criteria as listed in Table 6. No doubt *R. bengalensis* and *D. geloniodes* are potential candidates by virtue of the fact that both plants are strong hyperaccumulators of Ni. In terms of leave biomass (Table 2), *R. bengalensis* is comparable or bet-

Table 6:   Relative performance of R. bengalensis and D. gelonioides with respect to the criteria required for a candidate plant in phytoremediation and phytoming technologies

Criteria	Relative Performance	
	R. bengalensis	*G. geloniodes*
Metal tolerant	Excellent	Excellent
Strong metal hyperaccumulator	Excellent	Excellent
High above ground biomass	Good	Poor
Short life cycle	Poor	Poor
Fast growth rate	Poor	Poor
Easy to propagate	Poor	Poor

ter than other metal hyperaccumulators being investigated elsewhere for phytoremediation or phytomining. By contrast, biomass production of *D. geloniodes* is low. Like many other hyperaccumulators, both plant species unfortunately have relatively slow growth rate. These limiting factors could be potentially overcome by means of conventional breeding (Chaney et al., 2000) and genetic engineering (Kenlampi et al., 2000) approaches.

CONCLUSION

Among the diverse flora found in Malaysia, at least two plant species namely *R. bengelensis* and *D. gelonioides* are strong Ni hyperaccumulators with foliar Ni content exceeding 1% dry matter. Although both species, and notably *R. benglensis*, has many attributes that make them potential candidate plants for phytoremediation and phytomining, the viability of these technologies using these indigenous plants need to be studied further. At the same time since metal hyperaccumulating plants in this region are increasingly becoming endangered (as a consequence of development activities and natural disasters) it is crucial to sustain geobotanical explorations to identify and collect indigenous metal hyperaccumulators, and if possible establish a germplasm facility for future R & D.

ACKNOWLEDGEMENTS

Support from Universiti Malaysia Sabah is gratefully acknowledge. A very special gratitude also to the late Mr. Sukup Akin, whose botanical knowledge has helped us in finding the elusive field specimens of *R. bengalensis* during our numerous field work in Sabah, Malaysia.

REFERENCES

Anderson, C. W. N., Brooks, R. R., Chiarucci, A., LaCoste, C. J., Leblanc, M., Robinson, B. H, Simcok, R. and Stewart, R. B. (1999), *Phytomining of nickel, thallium and gold,* Journal of Geochemical Exploration 67, 407–415.

Baker, A. J. M., Proctor, J., van Balgooy, M. M. J. & Reeves, R. D. (1992), *Hyperaccumulation of nickel by the ultramafic flora of Palawan, Republic of the Philippines,* in: The Vegetation of Ultramafic (Serpentine) Soils. Proctor, J., Baker, A. J. M. & Reeves, R. D. (eds.), Intercept Ltd., Andover, Hants, U.K., 291–304.

Baker, A. J. M., McGrath, S. P., Sidoli, C. M. D., Reeves, R. D. (1994), *The possibility of in situ heavy metal decontamination of polluted soils using crops of metal-accumulating plants,* Resources, Conservation and Recycling 11, 41–49.

Baker, A. J. M., McGrath, S. P., Reeves, R. D. , Smith, J. A. C. (2000), *Metal hyperaccumulating plants: A review of the ecology and physiology of a biological resource for phytoremediation of metal polluted soils,* in: Terry, N., Banelos, G. (eds.), Phytoremediation of Contaminated Soil and Water, CRC Press Inc., Bota Raton, Fl, USA., 86–107.

Brooks, R. R., Chambers, M. F., Nicks, L. J., Robinson, B. H. (1998), *Phytomining,* Trends in Plant Sciences 3, 359–362.

Brooks, R. R. (1987), *Serpentine and its Vegetation,* Discoride Press, Beckenham, UK., 316–329.

Brooks, R. R., Wither, E. D. (1977), *Nickel hyperaccumulation of Rinorea bengalensis,* (Wall) O. K. Journal of Geochemical Exploration 7, 295–300.

Brooks, R. R. and Radford, C. C. (1978),
Nickel hyperaccumulation by European species of the genus Alyssum,
Proc. Roy. Soc. London Sec. B200, 217–224.

Brooks, R. R. & Robinson, B. H. (1998),
*The potential use of hyperaccumulators and other plants for
phytomining*, in: Brook, R.R. (ed.), Plants That Hyperaccumulate Heavy
Metals, CAB International, Walingford, UK., 327–356.

Brooks, R. R., Lee, J., Reeves, R. D., Jaffre, T. (1977a),
*Detection of nickeliferous rocks by analysis of herbarium specimens of
indicator plants*, Journal of Geochemical Exploration 7, 49–57.

Brooks, R. R., Wither, E. D., Zepernick, B. (1977b),
Cobalt and nickel in Rinorea species,
Plant and Soil 47, 707–712.

Brooks, R. R., Chambers, M. F., Nicks, L. J., Robinson, B. H. (1998),
Phytomining, Trends in Plant Sciences 3, 359–362.

Chaney, R. L., Angle, J. S., Baker, A. J. M., Li, Y. M. (1998),
Methods for phytomining of nickel, cobalt and other metals from soils,
US Patent 5711784.

Chaney, R. L., Li, Y.-M., Angle, J. S., Baker, A. J. M., Reeves, R. D.,
Brown, S. L., Homer, F. A., Malik, M., Chin, M. (2000),
*Improving metal hyperaccumulator wild plants to develop commercial
phytoextraction systems: approaches and progress*,
in: Terry, N. & Banelos, G. (eds.), Phytoremediation of Contaminated
Soil and Water, CRC Press Inc., Bota Raton, FL, USA., 129–158.

Jopony, M. and Baker, A. J. M. (2000),
Hyperaccumulation of nickel and zinc by plants of Southeast Asia,
Proceedings of International Conference on Soil Remediation
(SoilRem 2000), 15–19 October, 2000, Hangzhou, China.

Kenlampi, S., Schat, H., Vangronsveld, J., Verkleiji, J. A. C.,
van der Lelie, D.,Mergeay, H., and Tervahanta, A. I. (2000),
*Genetic engineering in the improvement of plants for
phytoremediation of metal polluted soils,*
Environmental Pollution 107, 225–231.

Jaffre, T., Schmid, M. (1974),
*Accumulation of nickel by a Rubiaceae of New Caledonia:
Psychotria douarrei (G. Beauvisage),*
C. R. Acad. Sci. Paris, Ser. D 278, 1727–1730.

Jaffre, T., Brooks, R. R., Lee, J. and Reeves, R. D. (1976),
Sebertia acuminata: a nickel accumulating plant from New Caledonia,
Science 193, 579–580.

Ng, F. S. P. (1995),
Violaceae. In Tree Flora of Malaya,
A Manual for Foresters.Vol. 3, Longman, Kuala Lumpur.

Nicks, L. J. and Chambers, M. F. (1998),
A pioneering study of the potential phytomining for nickel,
in: Brooks, R.R. (ed.), Plants That Hyperaccumulate Heavy Metals,
CAB International, Walingford, UK., 313–326.

Proctor, J., Phillipps, C., Duff, G. K., Heany, A., Robertson, F. N. (1989),
*Ecological studies on Gunung Silam, a small ultrabasic mountain in
Sabah,* Malaysia. Journal of Ecology 77, 317–331.

Raskin, I., Smith, R. D., Salt, D. E. (1997),
*Phytoremediation of metals: using plants to remove pollutants from the
environment,* Current Opinion in Biotechnology 8, 221–226.

Reeves, R. D. (2002),
*Tropical hyperaccumulators of metals and their potential for
phytoextraction,* Plant and Soil (in press).

Reeves, R. D. (1992),
Hyperaccumulation of nickel by serpentine plants, in: Proctor, J.,
Baker, A. J. M., Reeves, R. D. (eds), The Vegetation of Ultramafic
(Serpentine) Soils, Intercept Ltd., Andover, Hants, U.K., 253–277.

Reeves, R. D. and Brooks, R. R. (1983b),
*European species of Thlaspi, L. (Cruciferae) as indicators of nickel and
zinc*, Journal of Geochemical Exploration 18, 275–283.

Reeves, R. D. and Baker, A. J. M. (2000),
*Metal accumulating plants. In Phytoremediation of Toxic Metals:
Using Plants to Clean Up the Environment,*
in: Raskin, I., Ensley, B. (eds.), Wiley and Sons, New York., 193–229.

Robinson, B. H., Chiarucci, A., Brooks, R. R., Petit, D., Kirkman, J. H.,
Greg, P. E. H. and De Dominicis, V. (1997a),
*The nickel hyperaccumulator plant Alyssum bertolonii as a potential
agent for phytoremediation and phytomining of nickel,*
Journal Geochemical Exploration 59, 75–86.

Salt, D. E., Blaylock, M., Kumar, N. P. B. A., Dushenkov, V.,
Ensley, B. D., Chet, I. and Raskin, I. (1995),
*Phytoremediation: a novel strategy for the removal of toxic metals from
the environment using plants*, Biotechnology 13, 468–474.

Severne, B. C. and Brooks, R. R. (1972),
A nickel-accumulating plant from Western Australia, Planta 103, 91–94.

Visoottiviseth, P., Francesconi, K. and Sridokchan, W. (2002),
*The potential of Thai indiginous plant species for phytoremediation of
arsenic contaminated land*, Environmental Pollution 118, 453–461.

Wither, E. D. and Brooks, R. R. (1977),
Hyperaccumulation of nickel by some plants of Southeast Asia,
Journal of Geochemical Exploration 8, 579–583.